THE RIVERBONES

STUMBLING AFTER EDEN IN THE JUNGLES OF SURINAME

THE RIVERBONES

ANDREW WESTOLL

Emblem

Library and Archives Canada Cataloguing in Publication

Westoll, Andrew
    The riverbones : stumbling after Eden in the jungles of Suriname / Andrew Westoll.

ISBN 978-0-7710-8875-9

1. Westoll, Andrew – Travel – Suriname. 2. Suriname – Description and travel. 3. Jungle ecology – Suriname. 4. Human geography – Suriname. 5. Natural history – Suriname. I. Title.

F2413.W48 2008   918.8304'32   C2008-900938-X

We acknowledge the financial support of the Government of Canada through the Book Publishing Industry Development Program and that of the Government of Ontario through the Ontario Media Development Corporation's Ontario Book Initiative. We further acknowledge the support of the Canada Council for the Arts and the Ontario Arts Council for our publishing program.

Quotations from *First Time: The Historical Vision of an African American People* by Richard Price are reproduced by permission of the author.

Photographs by the author appear on pages 8, 37, 71, 110, 133, 267, 297, 330, 342, and 360. All other photographs are reproduced by permission of Jason Rothe.

Typeset in Electra by M&S, Toronto
Printed and bound in Canada

This book was produced using ancient-forest friendly papers.

McClelland & Stewart Ltd.
75 Sherbourne Street
Toronto, Ontario
M5A 2P9
www.mcclelland.com

1   2   3   4   5   12   11   10   09   08

For my parents, Maureen and Neil.
For my jungle grandfather, Petrus Tjappa.
For Jules, Julia, and the rest of the Raleighvallen monkeys.
Grantangi.

The man looks like the tree the man is killing.

EDUARDO GALEANO

How will we feel the end of nature?

BILL McKIBBEN

Wan bon	One tree
someni wiwiri	so many leaves
wan bon	one tree

ROBIN "DOBRU" RAVELES,
SURINAMESE NATIONALIST POET,
FROM HIS POEM WAN BON

CONTENTS

THE DYING ANIMAL

Tonight, on my way down the stairs from the cookhouse, a small bat slams into my hand. He drops straight to the ground and lands with a soft thud. I've always believed they would steer clear – a basic assumption of life in the rainforest is that bats will avoid you – but it turns out that's not true. I've thumped this one with my hand and now he's a dark stillness on the ground.

I turn on my headlamp and lean closer. He squirms away, pulling himself along the ground pitifully. He tries to fly but only makes it half a metre. This bat is going to die tonight. I have killed this animal and he's not even dead yet.

I follow him for an hour, from the clearing in front of the cookhouse into the dense jungle underbrush. I don't shine the light directly on him; I lose him a number of times. The humid air thickens as the forest wraps around us, the musty stench of decaying plants, the lowering canopy of twisted vines, the astonishing claustrophobia of innumerable whispering trees. To my left, the moonlight dapples a bank of delicate ferns. To my right, an impenetrable thicket of lianas dangles a metre or so off the ground. Underfoot, the soft floor of the jungle, the black soil made of everything dead or dying.

The bat slows, his movements become more pronounced. My guilty hand recalls a moment of silken fur. And then I remember walking with Emma in that other jungle, two weeks before I left

1

Toronto. We found a struggling pigeon on the sidewalk. At first I thought her wing was broken, but then her movements slowed. The Polish man with the broom tried to sweep her away, and I realized we were watching not a fallen body but a crashing life. She stopped moving and Emma let out a gasp – a whole, entire thing gone.

I searched my bag for a piece of paper, wrapped up the dead bird, and carried her into a nearby alley. I felt her warmth against my palm, passing through the paper, as if she were asking for protection even though she was long past needing it. I left her in a quiet corner, buried beneath a handful of soil and leaves, and returned to the street, where Emma called me her jungle hero – already, way back then. Just days before I returned to this unknown country for some unknown reason. Just days before I left.

On my favourite map of the world, the National Geographic's map of earth-borne light, Suriname barely exists. There is just one small point of light, the capital, Paramaribo, on the northeast shoulder of South America. In this photograph of the sleeping world taken from space, the rest of Suriname is dark – as if no one lives here and nothing happens in the dense jungle that lies to the north of Amazonia.

When I was twenty-three years old, I lived deep inside this darkness for a year. My home was a place called Raleighvallen, a remote outpost in the Central Suriname Nature Reserve, which is the largest protected tract of pristine rainforest on the planet. I was there to study monkeys.

I lived with six other researchers in a bare-bones building on the eastern bank of the Coppename River, a mile south of a small tourist facility called Foengoe Island. All of us were young, aspiring primatologists who couldn't believe our luck. We'd been given an opportunity thousands of biology graduates would have killed for –

an entire year in the middle of a virgin rainforest no one had heard of, surrounded by troops of brown capuchin monkeys no one had studied. We were connected to the rest of the world only by a wonky satellite phone and trips to the capital every two months.

We spent twelve hours a day, seven days a week, following troops of monkeys through an untouched wilderness. Our neighbours included all of South America's most poisonous snakes, the vicious fer-de-lance, the stubborn bushmaster. Jaguars roamed the forests at night and left eviscerated deer carcasses in the middle of our trails. We routinely crossed paths with herds of wild boar – some more than two hundred strong – that could trample a human in seconds.

We were the Jane Goodalls, the Dian Fosseys, the Biruté Galdikases of the New World. We wrote down, in obsessive detail, everything we saw: where the monkeys went, what they ate, who was friends with whom. We gave them names like Agnes and Wacky and Suri Rama and Mignon. They became our entire world, our reason for waking in the morning and our reason for collapsing into our bunks at night. Though well-fed we grew skinny, our cheekbones and ribcages announcing themselves, our fat reserves burning, our leg muscles shrinking, our bodies slowly evaporating. Physical and mental exhaustion became indistinguishable and a way of life. We lived on the edge of breakdowns and the known world. Sometimes we felt the monkeys were studying us.

When I finished my research contract I left Suriname right away. But the country stayed with me. I was hooked on the place. Over the next few years, I read every book about Suriname I could find, bookmarked every website that mentioned it and subscribed to online European newspapers I couldn't even read. I learned that Suriname was once a Dutch colony. I learned that it sits atop the Guiana Shield, a 1.7 billion-year-old sheet of Precambrian rock that encompasses more than two million square kilometres of tropical wilderness and is one of the oldest geological formations on the surface of the earth. I learned that for much of the Shield's existence, most modern tropical regions simply did not yet exist (such as Western Amazonia) or were covered over by shallow seas (such as Central America). I was struck by this and compiled a list of all the ancient, untouched places that have been hailed as the Last Eden – Patagonia in Argentina, the Okavango Delta of Botswana, the Ndoki region of the Democratic Republic of Congo, the wild island of Borneo – and added the jungles of Suriname to that list.

I collected obscure cultural references to the country the way a young hipster collects vintage T-shirts. In *The Silence of the Lambs*, Hannibal Lecter's shipment of Death's-head Hawkmoths comes from Suriname. In Gabriel García Márquez's *Love in the Time of Cholera*, a mischievous parrot from Paramaribo occasions the death, by fall from a ladder, of Dr. Urbino. In J.R.R. Tolkien's invented language of Quenya, the word *surinen* means "in the wind," but the Surinen were actually the Amerindians who furnished Suriname with its name. Rudd Gullit, my favourite footballer of all time, was born in Suriname, as were Edgar Davids, Clarence Seedorf, Patrick Kluivert, Frank Rijkaard, Jimmy Floyd Hasselbaink, and most other dark-skinned members of the Dutch national football team which I had supported, for no legitimate reason, since the age of eight. And, finally, during the last frenzied chapter of Don DeLillo's *White Noise*, protagonist Jack Gladney wonders if perhaps the man he's about to murder, the deranged Dylar dealer Willie Mink, might, in fact, be Surinamese. "Where was Surinam?" thinks Gladney, moments before he opens fire.

I shared this trivia with strangers on street corners, in elevators, at grocery stores, at dinner parties. I'd never been to Thailand or Peru or Prague, as it seemed everyone else my age had, and perhaps I was overcompensating for this. I told anyone who would listen about my monkeys, about what it was like to live in the largest undisturbed rainforest on earth. But I also told them stories about the rest of the country, things I'd only read about: mysterious Amerindian shamans, superstitious tribes of rebel African slaves, zealous Moravian missionaries, outlaw Brazilian *garimpeiros*, exhausted jungle gold mines, a fetid lake with the dead canopy of a drowned rainforest at its surface, the aftermath of a six-year civil war, and an unsolved murder mystery fired with political intrigue that continues to haunt the nation.

Don DeLillo was right. No one had heard of the place. Most thought it was a post-colonial fragment somewhere in Africa, a few figured Southeast Asia, and the rest thought I was making it up. Occasionally, these doubters came close to convincing me. After all, I'd accepted the job as a monkey researcher thinking I was headed somewhere near Vietnam. It was only when my airline tickets showed up that I looked up Suriname on a map and found, to my amazement, that there were monkeys in need of study at the bottom of the Caribbean.

Friends and family often asked why I was so haunted by a place where so little happens, where there are fewer people than in any other South American nation, where 90 per cent of the country is jungle, where deep in the bush there are only 0.2 people for every square kilometre, a density similar to Nunavut in the Canadian North. I said I didn't know. Suriname sounds like such a magical place that if I hadn't already been there I'd swear it was imaginary. It reminds me of Macondo, Márquez's fictional land.

Soon I ran out of new information about the place. With nothing more to read, I imagined what it would be like to return. My life began to revolve around this idea. I stopped making long-term plans. I re-read those books. I fell out of love with science and began writing. I wrote my first novel (a bad one). I tried, and failed, to get it published (and rightly so). I waited for a sign. I waited for time to guide me.

Finally, five years later, I am back in Suriname, but I'm no longer a primatologist. My office used to be the depths of the rainforest and now it's a thatched-roof power shed on Foengoe Island. Conservation International flew me down here. My job is to write travel brochures.

It's movie night in the jungle. The cookhouse is filled with workers. On one side sit the Kwinti Maroons, proud descendants of African slaves who escaped into the Surinamese rainforest three hundred years ago. Opposite the Kwinti are the timber framers, Western volunteers who've just arrived from Canada, America, and the U.K. to build a new tourist centre here.

Last week the film was *Buffalo Soldiers*; this week it's *Burden of Dreams*, the documentary about Werner Herzog's classic *Fitzcarraldo*, especially its most famous scene, when Fitzcarraldo brazenly hauls a 320-tonne steamship over a jungle mountain to reach untapped rubber trees. Obsessed with realism, Herzog refused to use mockups or studio sets for this. Instead, he travelled to Ecuador's rainforest, hired more than twelve hundred locals, bought a real steamship, and attempted to pull it over a mountain. His epic, four-year struggle to shoot the scene is now part of filmmaking legend.

Before the movie starts, Chuck Hutchinson addresses the audience. He is a senior director of conservation and tourism planning with Conservation International (CI), one of the world's most influential environmental non-governmental organizations (NGOs).

Chuck explains how the Central Suriname Nature Reserve (CSNR) was created in July 1998, after CI fought off Southeast Asian logging conglomerates and convinced the Surinamese government – with an initial donation of $1 million and the promise of further millions in trust – that the "existence value" of the jungle (the potential for ecotourism projects and pharmaceutical discovery) far

6

outweighed its "destruction value" (timber revenues). The CSNR encompasses sixteen thousand square kilometres of undisturbed tropical rainforest in the heart of the country – one-tenth of Suriname's land mass – and was recently named a UNESCO World Heritage Site.

When the CSNR was announced it was hailed as a staggering triumph for rainforest conservation. But for Chuck's boss, Russell Mittermeier, the triumph must have been especially sweet. Long before he became renowned as a Harvard-trained primatologist, biodiversity advocate, public relations whiz, *Time* magazine Hero of the Planet, and something of a modern-day Tarzan, Mittermeier conducted his Ph.D. research in these jungles. Working from a remote camp at the foot of the Voltzberg Dome – the highlight of a trip to Raleighvallen, a massive black-granite mountain that resembles a loaf of burnt sourdough puncturing the jungle canopy – he carried out the first study of Suriname's spider monkey. For Mittermeier, no place on earth is wilder, no place more worth saving. He was the spark behind the idea of the CSNR and has paid countless return visits to Raleighvallen. An iconic image of him appeared in *Time*, on top of the Voltzberg with his machete raised in triumph, a spectacular panorama of rainforest stretching all around him.

Chuck Hutchinson is Mittermeier's man-on-the-ground here, a regal, somewhat flamboyant man with a smoothly shaved scalp, a salt-and-pepper goatee, and a penchant for West African fashions. Chuck holds a gin and tonic as he speaks. With a few words and a sweep of his arm, he gives a short history lesson while illustrating the basic principles of ecological interconnectedness.

He describes how the waters of the Amazon carry silt from high in the Peruvian Andes to the river's mouth on the east coast of Brazil – thousands of kilometres and almost an entire continent away from Peru. From there the silt is driven north and west by the Atlantic

current, hugging the coastline of South America, and is eventually deposited on Suriname's shore. The only plants that grow in this Peruvian mud are mangroves. This is why Suriname's shoreline was dubbed the Wild Coast by Spanish explorers who first encountered it more than five hundred years ago. With its long tidal mudflats, tangled mangrove thickets, and mosquitoes the size of quarters, the region was deemed inhospitable by the conquistadores, who quickly moved on.

Chuck tells us that the tourism centre, the reason the timber framers are here, epitomizes the new conservation economy and gives hope for a sustainable future in South America. The Amazon rainforest might be a dying animal – in Brazil, a few hundred kilometres to the south, it is already far gone – but here, on the Guiana Shield, there is still hope. As the conservationist, the ecotourist, and humanity itself stumbles after Eden, Suriname serves as a quiet example of all we've got left to save.

This country has the potential to become one of the premier ecotourism destinations in the world, Chuck says, the next Costa Rica. But Suriname's relative obscurity will not last forever. We need to create a brand around the forest in order to protect it into the future, he says. The new visitor's pavilion will do just that.

Our movie screen is a white sheet hanging from the rafters in the middle of the room. Most of the Kwinti are on the far side of the screen, watching a mirror image of the film. They don't seem to care. The opening sequence is shot from an airplane, a chocolate river snaking through thick jungle near the Ecuadorian border with Peru. It could have been shot on the flight that brought me here three days ago.

As fist-sized moths swarm the projector, we watch Werner Herzog slowly lose his mind. We watch him struggle to pull his steamship up a remote jungle hillside against impossible odds. Mick Jagger pulls out of the film to go on tour with the Rolling Stones. Jason Robards pulls out with a horrible case of dysentery. Klaus Kinski is called in as a last-minute replacement. Moving the steamship is made impossible by the longest dry season in recorded history. The film's backers in Germany threaten to pull out. Herzog's crew of Amerindians staves off boredom by launching deadly raids on neighbouring villages. The filmmaker hires prostitutes to service his restless workers and keep them calm. The bulldozer doesn't work. The camp sanitation system overflows. The engineer quits. The only soccer ball in camp has a hole in it. "I am running out of fantasy," says Herzog. But still he continues. "If I abandon this project I would be a man without dreams. I live my life or I end my life with this project."

The timber framers are riveted. Chuck's selection was a good one – the magnitude of their mission here is slowly sinking in. The Kwinti, on the other hand, don't speak English and are missing the point. They laugh when the winching apparatus fails and the boat slips back down the hill, almost killing a man. They only become serious when they see footage of an Indian trying to kill a snake.

I watch from outside the cookhouse with the Kwinti who couldn't get a seat. Bigga-man is here. So are Set-man, Soldier, and Doctor Five. Every few minutes, a leafcutter ant chomps my big toe with its massive pincers and I calmly rip it off. Behind us, hundreds of them are scurrying back to their nest along a well-worn pathway.

"This will be one of the last feature films with authentic Indians in it," says Herzog. "We are losing them, their languages, their individualities, their mythologies. Soon we'll just be cities, with skyscrapers, with one culture. The American culture."

As things get worse, Herzog turns his wrath from mankind to the jungle itself: "Kinski always says it's full of erotic elements. I don't see so much erotica, I see it more full of obscenity. It's just – nature here is vile and base. I wouldn't see anything erotic here. I would see fornication and asphyxiation and choking and fighting for survival and growing and just rotting away. Of course there's a lot of misery, but it is the same misery that is all around us. The trees here are in misery and the birds are in misery. I don't think they sing. They just screech in pain."

The timber framers are laughing with the Maroons now. Herzog is losing it. But I am standing in the same spot where I killed the bat and am transfixed by what Herzog is saying.

"It's like a curse weighing on an entire landscape. And whoever goes too deep into this has his share of that curse. So we are cursed with what we are doing here. . . . There is no harmony in the universe. We have to get acquainted to this idea. That there is no real harmony as we have conceived it. But, when I say this, I say this full of admiration for the jungle. It is not that I hate it. I love it. I love it very much. But I love it against my better judgment."

The movie ends with Herzog's steamship cresting the hill. The cookhouse fills with applause. Writing about the ordeal twenty years

10

later, the filmmaker would mock himself as the "Conquistador of the Useless."

I spend my days writing on my laptop computer. Sometimes I am joined by a monkey researcher, one of the new crew, the young graduates who are living my old life. They come to the island on their days off to use the internet and poke fun at the small groups of Dutch tourists.

They tell me about my monkeys, the troop we nicknamed the Starbucks, how Jules was kicked out after a vicious fight with Bruce, how Banana had her first baby, how the resident harpy eagle hasn't killed any capuchins yet but is often seen with a baby red howler in her talons. They ask me when I'm coming to visit and I tell them soon, when I can find the time and a boatman willing to take me. And then they leave the island in their little camp boat, paddling upriver and away from the world, and I am left to swelter with jealousy in my shed.

Although it was an incredible adventure, after a month of following the monkeys I began to feel I was missing something. I felt my task of reducing the actions of the monkeys down to quantifiable datapoints was distancing me from a richer experience. Wandering with my professor and listening to her rattle off the Latin names of every plant and animal began to drive me mad. The satisfaction she derived from nomenclature was, to my eyes, akin to the satisfaction humans derive from owning and controlling the natural world. The more Latin I learned, the more closed off I felt from the jungle.

Meanwhile, our little camp library had become a war zone – primate journals and books on chimpanzee politics battling it out with Tolstoy, Michaels, Ford, and García Márquez. I soon found myself torn between the scientific "truths" we were discovering in the forest every day and the broader "truths" offered by the great writers I admired.

So I began faking the data. Whenever I was alone I would sit beneath the monkeys and write down random numbers without even looking up. Group scans, canopy heights, troop dispersals, foods eaten. I would sit beneath a tree with an unpronounceable Latin name while monkeys gorged themselves among the branches, the husks and rinds and seeds showering down around me, and I wouldn't write it down. I just fucking refused. I quietly sabotaged the project one day at a time. Monkeys would shit and I wouldn't collect it – priceless DNA disappearing straight into the underbrush. Monkeys would scream at each other and I wouldn't record it – important social hierarchy data dissolving in the humid air. Instead, I scribbled plotlines and character sketches on yellow data paper when I wasn't swatting at enormous black-red wasps that hovered around my body, stalking giant armadillos I'd seen disappearing into their burrows, or waiting for the next forest turtle to rumble past. I loved my monkeys, I loved being in the forest. But I hated the work. I'd been set loose in a jungle paradise and all I cared about was the hot dripping mess that clawed at my body, the steaming world that stung my palms when I reached out to it, the poisonous spines of urticaceous caterpillars on the undersides of leaves.

My rebellion lasted about a week. Then I snapped out of it and became a professional again. And now the same things that pushed me from science – the unbridled rationality, the deconstruction of mystery, the arrogance of ownership that comes with classification – are what I'm sensing at Raleighvallen.

Ecotourism here looks like a new form of imperialism. The new visitor's centre, financed with money from Wal-Mart and Intel, will dwarf the very forest it was built to promote. Soon it will host an internet café, crowds of Dutch tourists drinking Coca-Cola, hordes of European teenagers performing cannonballs into the water and getting high on the rocks, gaggles of young women on vacation looking to bed a wild, primitive Maroon. Chuck says this is what it takes to preserve things in the new conservation economy – that the "existence value" of a jungle must be greater than its "destruction value" before it can be saved – but something is lost when we look at the natural world this way. Something invisible and ethereal but no less a force.

I can't escape the feeling that the spirits of Raleighvallen are shifting, that mother nature is once again on the run. I came here to find closure, but all I've found is what looks like the beginning of the end.

One hot morning, Petrus Tjappa appears in the doorway of the power shed, an enormous chainsaw balanced on his head.

"*San?*" he says, confused. I am the first monkey researcher to ever return to Raleighvallen.

Petrus has lived on Foengoe Island for more than thirty years. He works for STINASU, the Surinamese conservation and tourism agency, and is generally considered the patriarch of Raleighvallen. No one knows how old he is because he can't remember his birthday. Most guesses put him around sixty-five years old.

Five years ago, Petrus was our jungle grandfather. He came to see us every day, travelling upriver at dusk to make sure we'd all returned safely from the bush. He'd bring us Tupperware containers filled with fried *anyumara*, a delicious rich river fish, the oily flesh falling from the bones, the skin charred and spiced with *fonfon* pepper. We'd invite him in for dinner but he'd always refuse; he was simply happy to have found us all safe and sound. We'd stand in the doorway and chat about what we'd seen that day, tell him about the monkeys, and show him the latest video footage of the monkeys bashing *phenakospermum guyannense* nuts on the branch forks of trees. Sometimes we'd give Petrus ibuprofen for his chronic stomach ache, or a glass of river water sweetened with cherry *stroop*. Then we'd walk him back down to his boat with our flashlights, the sun long gone, tropical lightning flickering on the horizon. He was everyone's favourite islander, the only man who spoke slowly enough for us to learn Sranantongo, Suriname's lingua franca. The only man who seemed genuinely concerned for us, the strange kids from the West living in the jungle. He was a part of our family. When we left after a year, he gave us carvings he'd made and asked us for our addresses. In return I gave him my birthday, April 22, Earth Day, and told him he could share it with me.

Petrus has just returned from a vacation in Witagron, three hours downriver. He wears the same pink Denver Nuggets ball-cap he wore five years ago. It's almost white now, bleached by the equatorial sun.

I ask Petrus how many children he has now. He gazes up at the sky with a serious look, his long wrinkled fingers emerging slowly from his fist as he counts. He holds up one hand, five fingers splayed.

"*Kande achttien*," he says. Eighteen children, maybe more.

During my year as a monkey researcher, our jungle home fell under a strange sort of spell. Friendships between us all began to fall apart. The social dynamics in our house became more and more complicated, fuelled by gossip and rumour and jealousy, and soon no one knew who they could trust. Then one night Petrus came over and asked to come inside. He sat us down around the dinner table and began to speak. I was able to catch a few phrases repeated over and over: *libi makandra* (live together), *no breiti* (unhappy), *ogri* (bad, evil). I gathered that Foengoe Island had been afflicted with a similar social ill. All of the wives had been sent downriver to their respective villages as a means of restoring the peace.

Petrus told us the chiefs from Witagron and Kaaimanston had visited Foengoe recently. They had travelled upstream to conduct a cleansing ceremony at Moedervallen, Mother Falls, the spiritual centre of Raleighvallen. The chiefs had heard about the trouble on the island and wanted to put an end to it, to spare their villages from any heartache or discontent that may have leaked into the waters of the river we all shared. All the men on the island had painted their naked bodies in white chalk, the colour of supplication, and participated in a ritual the day before.

Petrus motioned for us to follow him outside. He walked to the edge of the clearing where our sacred palm tree stood, a collection of empty liquor bottles at its base. Petrus had blessed this tree two years earlier, when the monkey project first began, asking the forest spirits to look after us. He bent down, removed the branches and debris that cluttered the makeshift shrine, and straightened a few of the bottles that had been knocked over. Then he pulled a *jugo*, a litre of Parbo beer, from his knapsack and cracked it open.

Petrus began to chant. He spoke softly, bent over with his eyes closed, his voice rising and falling. As he chanted, he scattered beer over the tops of the other bottles. His stance was submissive, reverent, and although I couldn't understand a word of it, something told me that this was not a prayer for protection – this was a prayer of apology. Petrus was trying to convince the forest spirits that, contrary to the way we'd been treating each other, we were worthy of living in this jungle cathedral.

When he was finished he looked exhausted, as if a huge weight had been slung over his shoulders. He'd just taken responsibility, in the eyes of the spirits, for all of our evils. Petrus passed the bottle around and told us to drink, and as he took what was left and rubbed it across his cheeks we all began to weep with shame.

I follow Petrus to his house. We pass *Tamandua*, the oldest tourist lodge on the island, built on the foundations of an even older lodge that was burned to the ground during the civil war that racked the country in the late 1980s. A decade earlier, Raleighvallen was a world-renowned paradise for tropical birdwatchers. Avian enthusiasts would flock here to add the crimson fruitcrow, the harpy eagle, and the spectacular Guianan cock-of-the-rock to their lifelong sightings lists. But with the military coup in 1980 and the war that followed six years later, Suriname's budding tourism industry crumbled along with the country's infrastructure. Power lines used to deliver electricity to Raleighvallen straight from Paramaribo, but now the only power comes from an array of solar panels that CI paid for. The airstrip, built in the 1960s, was destroyed by the Jungle Commando. After the war, all that remained of Raleighvallen were a few overgrown and charred concrete slabs on the banks of the Coppename.

Petrus invites me inside his house, the one he had to rebuild after the war, a tiny, thatched-roof hut beneath a majestic mango tree. Behind his house, Bigga-man and Doctor Five sit on a bench next to the massive tree trunk, holding plastic plates up to their chins and devouring slippery pieces of fish. I greet them – "*Morgu, morgu.*" – but all I get is a thumbs-up from Bigga and a grunt from Five. I leave my sandals on Petrus's rickety porch and step inside.

Petrus offers me a chair and disappears into his bedroom. He returns with three mouldy photo albums and flips through them slowly, his glasses on the end of his nose, his face twisted in concentration. At each new photo he pauses and rubs a finger over the plastic sleeve that houses it – these three albums hold the entire record of his life. I see black and white pictures of him as a young boy, surrounded by twenty of his family members, Paramaka Maroons with tough, smiling faces, all of them dressed in traditional clothing – overflowing women wrapped in *pangis*, proud men in *banya-koósus* and *kamisas*. Photos of all his children and grand-

16

children, the names of whom he's completely forgotten. A portrait of the young Petrus in a crisp STINASU uniform, a tan safari shirt with the Surinamese flag on the left breast.

He flips through recent colour photos, groups of young white people. The same blonde woman is in each of these shots: the professor, my old boss, Dr. Sue Boinski. Then we come upon a group I recognize. There I am at the back, tall and rail thin, my hair dyed blonde, the woven *mataklapu* Petrus gave me for Christmas held across my chest like a trophy.

Petrus opens the last album and stops at the first page. It's a picture of about thirty people standing in a dangerously overloaded dugout canoe. Everyone looks drunk and happy and their mouths are wide open in song. Petrus tells me the photo was taken two years ago in Nason, the village of his birth, six hours up the Marowijne River on the border with French Guiana. The party was a wake for his mother. It happens every year on Boxing Day. You should come this year, he says. I tell him it would be an honour.

I ask Petrus if he can take me upriver to Monkey Camp. He says no, he's sorry, the tourists are keeping him too busy. Then he asks me if I have a wife. I lie and say yes.

Emma and I first met a year ago in Montreal, at a dinner party in Outremont. She was a young novelist and I wanted to become one, so our friends thought we should meet. I remember her arrival. It was late fall, so she wore a scarf over her face and all I could see were her eyes. They were yellow and haunting, the eyes of a lion or perhaps the eyes of a wolf. Emma unravelled her scarf while her boyfriend said something implausibly stupid. Then she introduced herself and waltzed right past me.

Over dinner, I don't remember what anyone said because the first thing Emma did was tell us where her mother tongue came from, up over some range of hills with a scrappy crew of gypsies. For the rest of dinner I just wondered where she had come from and if perhaps she'd come over a hill I didn't know was in my immediate future. And maybe she was hardscrabble too, like the gypsies, and spoke a bit of their language and knew how to cook stew.

At one point, I went outside for a smoke and paced the broken-down balcony, watching Emma through the window as she held forth over the table of empty wine bottles. Because her boyfriend was gone, she could finally be herself. And I'd spent the last few years feeling nothing for various women. I was so tired of feeling nothing.

So I asked for her address. Then I returned to Vancouver and we began writing letters in which we told each other everything. She described in harrowing detail the abandonment she'd suffered, the wounds she continued to lick. I admitted my fixation with Suriname, the plans I'd made to return. We complained about the overbearing people we were dating, people who yearned to cut chunks out of our sides and store them in jars. We joked that someday we'd rescue each other, that someday we'd live in the same city.

Over time, our letters became love letters. Then, just weeks before I left Canada, Emma flew to Vancouver on a whim. It was only the second time we'd seen each other. We went camping at Alouette Lake.

The waters were like a drug. We felt high as kites after swimming, as we read our manuscripts out loud on the beach. I remember the exact moment of infatuation: she'd just bitten into a pear and I'd become hopelessly transfixed by the beautiful line of her jaw. Over her shoulder, miles away, the scar of a clear-cut. To distract myself I told her about the British Columbian trees, about everything we're losing. She thanked me for cheering her up.

We made love on the forest floor, between tourist trails on a bed of pine needles and moss. We became a part of North America's dwindling west-coast jungle, the trees that used to spread from

Alaska to Baja. We'd found that quiet garden where everything is young, reborn, and blooming.

Emma flew back to Toronto and I packed up my life. Then I followed her east. For the next two weeks, I couldn't make love to her deeply enough. Every tilt of her head or blink of her eye sent me reeling, and every word we shared, no matter how small, became weighted with thrilling consequence. I fell in love with Emma, became wholly obsessed by her. I had never before felt this strongly about a woman.

But during those last few days in Toronto I never questioned that I was leaving for Suriname. Fortunately, mercifully, neither did she. Instead, we scoured the newspaper for future apartments, planned future trips to Caribbean islands, agreed to become each other's first reader. I gave her a ring, which she slid onto her finger, and we committed ourselves – all she asked, through her tears at the airport, was that I avoid being eaten by jungle monsters.

I would be gone for a few months, and Emma would finish her second book. I would finally find closure with this country.

On that map of earth-borne light, there is hardly a single point of black on the eastern seaboard of the United States. It is awash with light, blinding even from space, and at the centre of this white is the island of Manhattan. This is in stark contrast to the black of Suriname. In 1667, a camera in space would have recorded no earth-borne light. That was the year the second Anglo-Dutch War came to an end, and the territory of Suriname was traded to the Dutch in return for the island of Manhattan.

The land the Dutch took over was little more than thick jungle and dark rivers, a land inhabited by animist tribes of Carib, Arowak, and Trio Indians. The Dutch quickly built Paramaribo on the Suriname River and established coffee and sugar plantations along the rivers closest to town. The Dutch West India Company shipped tens of thousands of captured West and Central Africans to toil as slaves on the plantations of Dutch Guiana, as Suriname was then known. Soon, Paramaribo became known throughout the New World as the "wooden city of the tropics," where Dutch planters lived like kings.

The Netherlands' slave owners were renowned for their cruelty, and the slaves of Suriname faced unimaginable suffering. Slaves who escaped and were later captured had their Achilles tendons cut. If they tried again, their right legs were amputated. Punishments included castration, roasting alive, and the "Spanish Whip." In 1759, when Voltaire needed a location for his satire of New World slavery he sent Candide to Suriname. On the road to Paramaribo, his protagonist stumbles upon a horribly deformed slave lying in the middle of the road. This is where the disillusioned world traveller formally renounces his belief in Panglossian Optimism.

"Yes Sir," said the slave, "it is the custom. . . . When we work in the sugar mills and we catch our finger in the millstone, they cut off our hand; when we try to run away, they cut off a leg; both things have happened to me. It is at this price that you eat sugar in Europe."

Even with the threat of these tortures, the interior of Dutch Guiana provided too tantalizing a glimpse of freedom for the slaves. In 1690, the colony's first slave revolt took place at a plantation on the Cassewinica Creek. Rebellion spread through the plantations, and scores of slaves escaped into the jungle under the cover of night.

The majority of rebels fled south. From makeshift villages deep in the bush, the strongest among them launched a guerrilla offensive against the Dutch, returning to the plantations at night to steal machetes and food, to set buildings ablaze, to murder particularly sadistic planters, and, most importantly, to whisk more of their people south to safety. When word of these raids reached Paramaribo, the Dutch launched countless war parties into the bush. They razed rebel villages, burned their gardens, and either killed slaves on the spot or dragged them back to town to face punishment.

The most famous account of Dutch Guiana in those days – and, to this day, one of the few book-length English travelogues about the country – was written in 1790 by Captain John Gabriel Stedman, an officer in the Scots Brigade in the Service of Holland. His *Narrative of a Five Years Expedition against the Revolted Negroes of Surinam in Guiana on the Wild Coast of South-America* describes the years he spent chasing down and killing fugitive slaves. Despite his occupation, Stedman offers a surprisingly candid portrayal of the cruelty of Dutch plantation owners. His moralizing on the institution of slavery, his love affair with a young slave woman named Joanna, and the book's chilling lithographs depicting the physical horrors suffered by the slaves (many of which were engraved by Stedman's friend, William Blake) made Stedman's *Narrative* a controversial publishing sensation across Europe.

The Dutch army had no chance of putting down the slave revolt. The soldiers had no experience in jungle combat; they'd been sent to fight for *patria* but ended up fighting for their lives. The rebels, meanwhile, were jungle people. Their ancestors had lived for centuries deep inside the towering rainforests of West and Central Africa. Their culture was geared to survival in the bush, their religion was based on forest spirits, both vengeful and benevolent. This was a time of epic battles and valiant deeds, of bows and arrows versus rifles and attack dogs, of secret agents and turncoat slaves – a time when "hunting" meant killing the Dutch.

The runaway slaves who survived men like Stedman eventually won amnesty from the Netherlands – the two largest groups had signed peace treaties before Stedman even arrived, nearly a century before the emancipation of slaves in Suriname. But the rebels remained in the small villages they'd founded on the Coppename, the Saramacca, the Marowijne, and the Suriname rivers. They survived as hunter-gardeners and created new languages spiced with English, Dutch, Portuguese, and Amerindian words. Over time, the rebel population gradually diverged into six distinct tribes – the Saramaka, Ndyuka, Kwinti, Matawai, Paramaka, and Aluku – each with its own rituals, customs, music, and language.

But these people were no longer African; they had become something more. The rebels had created a new African-American culture imbued with an entirely original creation story of hardship, horror, rebellion, and escape. The ancestors to whom they prayed were now the slaves who had fought for, and won, their freedom. They have been called by many names – Bush Negro, Bushland Creole, Bush Afro-American – but today they are known proudly as Maroon, a word derived from the Spanish *cimarrón*, connoting "wild and untamed."

Another week has passed. The moon is waxing and I wander the island, unable to sleep. A muffled noise like an AM radio draws me to the rear of the village, past an ancient cashew tree where the buildings back onto the forest.

I find Gordon, Shike, and a kid who calls himself Makelele – after the French soccer star – in the farthest hut. These three are among the youngest Kwinti on the island, just out of their teens. The room is thick with hash smoke and a small tape machine plays a live Surinamese rap record. The album sounds like it was recorded in a cave, the rapper's voice sounding hollow against the hum of the tape.

The boys welcome me with a knowing nod, clear a space on the bench, and Gordon rolls me a joint. They are talking about sex. Gordon is defending his dislike for certain acts, and the other two are teasing him mercilessly.

Gordon knows a little English and explains himself. "It is my culture," he says. "Tonight I wash in river. Every night, wash in river. It is medicine to stay clean. Life in jungle, sickness everywhere. With woman, you must stay clean."

"But it's their culture, too," I say, pointing to the others.

"Yes," says Gordon. "But their mother raised them badly."

Shike grips Gordon in a headlock. Gordon rolls his eyes.

"*Sa yu wani*," yells Shike into Gordon's ear. "*Sa yu wani*." Shike is a diehard Rastafarian, with thick dreads down to his waist. His pacifist philosophy is offset by his enormously muscled physique.

Suddenly, all three are rapping along to the record. Shike punches his free arm in the air and Makelele struts around the room like a drunken gangster. Gordon, still imprisoned, mouths the words into Shike's massive chest.

"*Suma disi?*" I ask. Who is this?

"Papa Touwtjie," says Gordon, finally breaking free. "Best rap man in Suriname."

"Dead," says Shike, perhaps the only word of English he knows.

"They shoot him," continues Gordon. "Two times. One leg, then other leg. First time, he live. Next time, he die. No rap man like Papa T."

Makelele and Shike are both looking down at Makelele's waist, where he is turning something slowly in his hand. Shike motions for me to come closer. As I approach I see what looks like a mis-shapen cassava tuber, the gnarled, starch-filled root that is the staple of the Maroon diet. Then I realize it's not a cassava, that it's Makelele's penis.

At first I think he wants my advice, my amateur medical opinion. His penis is riddled with bumps that look painfully swollen. But he looks up at me and smiles. So does Shike, and I realize this is nothing more than a show, a stoned display of masculine pride. He has small ball bearings beneath the skin of his penis. It is a rite of passage for young Maroon men, which they perform themselves with a sharp knife and perhaps a bottle of *palum*, the local rum. The marbles are called *bugaru*, and they provide a boost to the ego and sexual pleasure for them and their women.

The boys laugh raucously at the amazement on my face. I count eight *bugaru* on Makelele's penis before he tucks it back into his shorts. Gordon leans in to me. "Believe," he says, blowing hash smoke right in my face. "*Mi lobi* sex, too."

When I finally collapse into my hammock, countless shots of cheap cognac and two joints later, my head is spinning with kaleidoscopic visions. The forest spirits here are under siege, I am sure of it. Perhaps it began five hundred years ago, when the Wild Coast was first discovered by the Western world. Perhaps it has nothing to do with this new building. All I know is that image of Herzog's steamship being dragged up a jungle mountain is the most hauntingly unnatural thing I've ever seen. Whatever I'm searching for can no longer be found in Raleighvallen.

The next morning, incapacitated by a raging headache, I watch from my hammock as the timber framers swarm around their nearly finished building. A brilliant blue and green hummingbird – a fork-tailed woodnymph – alights on a feeder. The bird leaves and comes back, leaves and comes back, sampling a little bit of everything and returning to whatever seems richest. Right now it's the feeder but tomorrow it might be the flowers of the starfruit tree. It doesn't matter to the bird. And it doesn't matter to the traveller. Guide or no guide, we go wherever we feel we should go.

In two hours, a plane will arrive from Paramaribo with thirty-five tourists and then fly back empty. I tell Chuck I'm leaving.

Surprisingly, he understands. He tells me to be safe, to report back on what I find. The travel brochures can wait.

I pack my bags, then hike out to the south end of the island and down the steep hill to the river. The water is lower than usual, the rainy season long overdue, but in places it's still more than five metres deep. Schools of minnows and foot-long fish scatter as I step into the shallows. Five years ago, I never had the courage to do this – the Coppename is filled with piranha and caiman and electric eels. But now my priorities have changed. I don't have time to see my monkeys but at least I'll see my old home. I slip into the water and swim for it.

A half-hour later, I surface at the small beach where I used to bathe and walk into the jungle. The sun disappears and the air grows heavy. Thousands of insects begin to trill and I have to cover my ears. The trunks of great trees thrust up from buttress roots painted with rainbows of fungus, while above me, the canopy is earthen green, tightly wound with epiphytes, creepers, and twisted lianas. The ground is thick with rotting leaves and branches, the memories of previous lives, and drifting throughout is the breath of the jungle, that primordial stench of must and decay that crawls beneath your skin if you live here too long.

I hike up a short hill and emerge into a wide clearing, where a lone building stands in the blazing sun. Monkey Camp. My old home. I knock on the door. No one answers, so I pull the latch and step inside.

Data and gear are strewn everywhere – hand-drawn maps, charging batteries, rusted machetes, mud-crusted boots, cans of ineffectual bug dope, photos of long-dead monkeys, primatology conference posters with my name on them, a jar with an entire sloth skeleton inside. The monkey skull I found late one night while hiking home, the one I kicked across the trail before I realized what it was. I peer through the kitchen window at the water tower with its pipe going down the hill to the river, the enormous tree that cracked and toppled one day as I stood petrified at the sink, now half eaten, a pillar of decay lying on the ground. My eyes wander to the old propane stove, the Christmas decorations we made five years ago, swans and snowflakes from bright white paper, the small bloodstain on the floor that we could never get rid of. The long list of past field assistants in permanent black ink, again my name among them.

I revisit the bookshelf, our little camp library, the war zone where science finally lost me. I wander the rooms like a ghost, leave wet footprints on the cement floor and watch them disappear in the heat.

I leave a note for the researchers: "Hey Monkey People. I was here. Love, Andrew."

And then I leave.

On the way back to the river, I pass the old trailhead, the slim path that leads deeper into the jungle, that leads to my monkeys. Five years ago, every day was an adventure, and every day began right here. But those journeys are over now. A new journey is about to begin.

I swim downriver to Foengoe. The water is colder, and my breath comes harder, in heaving gasps. I have to stop a couple of times as my hangover rears its ugly head. Then I pull myself around a half-submerged boulder and come face to face with an angry river otter.

She is snorting territorially, diving and popping back up, circling in front of me as if wanting to spar. I climb onto the boulder and watch her, try to figure out if she is a regular otter or the vicious giant otter, the two-metre-long beasts you don't want to swim with. My eyes are not good. I can't tell if she has the characteristic white patch on her neck that would convince me to stay on this rock all night.

Thirty minutes pass. The otter keeps snorting and staring me down. I wait. Then, as quickly as she appeared she is gone, through the rocks to the other side of the channel. I listen for a few minutes but hear nothing. Then I slip back into the water and swim straight for those rocks because that's where I left my sandals.

As I approach I hear the twittering of young things among the stones. I stop swimming, tread water, catch my breath, and listen. An otter den filled with babies. Everything in the world worth protecting. I slide past, holding onto the rocks to keep quiet, and kick out into the home stretch, the current strong but moving with me, the water smooth and black through my fingers.

Up on the airstrip, alone and waiting for my plane to the city, I watch hundreds of grasshoppers rise from the flattened grass and fall like anxious angels.

PART ONE

The Converting World

"What we will have is what we ourselves will have built;
what we will value is what we ourselves can defend;
what we shall get is what we will wrest for ourselves;
what we shall be is what we will have made of ourselves."

DESI BOUTERSE, CHRISTMAS EVE, 1982

"Thirty years after its independence, Suriname is still like a hand
with five fingers that cannot make one fist."

DE WARE TIJD EDITORIAL, NOVEMBER 24, 2005

1

—

SREFIDENSI

I wake to the voices of a Moravian choir.

It's seven o'clock on November 25 and the staff of the Stadszending guesthouse, my home in Paramaribo, have gathered for bible study. They sing in Dutch and English; I pick out the occasional *Hallelujah* and *Master*. I crawl from my bed and drag myself to an internet café.

I send Emma a series of excited letters about my plans. A few hours south of Paramaribo lies one of the largest man-made lakes on the planet. Created in the early 1960s, the Brokopondo reservoir was supposed to usher in a new era of hope and prosperity for Suriname. Over time though (or so I've read), the lake has come to symbolize the suffering of Suriname's recent and distant past. I have spent the last five years reading books and surfing websites, trying to uncover Suriname's soul from a distance; now I will attempt to find it in person, starting at the lake.

Before I head south, there is a party to attend. Thirty years ago today, in 1975, Suriname became the youngest sovereign nation in South America. Today is *Srefidensi*, Independence Day, the biggest celebration of the year.

I wander through the Old Town, the original, seventeenth-century Dutch plantation town of three- and four-storey buildings built with tropical hardwood but sporting gabled roofs, white-washed façades, red-brick siding, and dark-green doors and shutters.

The ubiquitous red brick was imported from the Netherlands as the ballast in countless slave ships. The symmetry of the buildings is classically Old Dutch and trumpets wealth, but now the rooflines and gables sag with the weight of time.

Old Paramaribo, the Wooden City of the Tropics, is a UNESCO World Heritage Site, one of the finest exhibits of colonial settlement in the New World. More than 240 buildings in Paramaribo are protected and their interiors preserved, many of them former homes of prominent Dutch planters and merchants. These are now inhabited by travel agencies, bookstores, pharmacies, and internet cafés, bakeries selling almond pastries and *ollie bollen*, oily bread – Conservation International has its headquarters in one such building, as does the Canadian Consulate. But business inhabits these places without the usual spectacle. Many of these buildings are rumoured to have their original stone staircases, wood floors, rococo paintings, and opulent decor. For the buildings that survived two massive fires that swept these streets in the 1800s, little has changed in more than three hundred years.

On the Waterkant, it is easy to imagine the scene three centuries ago – slave ships docked just offshore, ready to trade their human cargo for sugar and coffee bound for Europe; tent-boats paddled by slaves sold to plantations on the Suriname, Commewijne, and Cottica rivers. The cobblestone street bustling with horses and carts, bellowing merchants, and Dutch women out for a stroll under their parasols, dark-skinned servants in tow.

I follow the roar of a loudspeaker along the Waterkant to Independence Square, the heart of Paramaribo, where I find thousands of people – Creoles, Javanese, Hindustani, Chinese, the occasional whites, all the colours of Suriname – dressed in the red, yellow, green, and white of their flag, and baking in the sun. Children hang from the statues of Johan Adolf Pengel and Jaggernath Lachmon, two of Suriname's revered old leaders, wreaths of orange flowers around their necks.

After slavery was abolished in 1863, the Dutch imported thousands of indentured labourers from Indonesia, India, and China to keep the plantations running. Their descendants remained in the colony and today make up the majority of its population. The minority by a slim margin is Afro-Surinamese – the Maroons who live in the jungle and the Creoles who live in the city. Surinamese society is a bustling mixture of cultures, heavily influenced by centuries of Dutch rule, less so by smaller groups of Jews, Brazilians, and indigenous Amerindians. The official language is Dutch but most casual conversation is conducted in Sranantongo, which most people speak.

Suriname also inherited the tradition of religious tolerance from the Dutch. Paramaribo is dotted with Islamic mosques, Hindu mandiras, Jewish synagogues, and Catholic churches. Paramaribo may be the only place on earth where a synagogue and a mosque sit side by side on Keizerstraat; their congregations share a parking lot. Visitors to the town are amazed to find a city of only 230,000 people where more than twenty languages are spoken, where every major religion is freely practised. For this alone, modern Suriname is often lauded as the world's finest example of a peaceful, plural society.

I join the throngs crowding the square. On the lawn in the centre, where Captain John Gabriel Stedman once trained his troops, regiments from Suriname's army, navy, and police force stand at attention with their counterparts from Venezuela, Guyana, and French Guiana. They bear stoic smiles and machine guns over their shoulders. Opposite them is a makeshift bandstand where a coterie of government officials and foreign dignitaries sit fanning themselves in front of the magnificent Presidential Palace. The Palace balcony is crowded with well-heeled well-wishers, the façade above them adorned with the Society of Suriname coat of arms, the pillars below stark and white. To their left is Fort Zeelandia, the old military barracks where rebellious slaves were tortured and slain hundreds of years ago, its haunted halls and cellars now graced by the national museum.

A military band strikes up and the soldiers leave the square to parade through the streets. The people cheer when the dignitaries step down from the bandstand and make their way to the Palace. As the ladies and gentlemen slip through a side gate, I glimpse the makings of a lavish party.

"What's going on in there?" I ask an old Creole, his face riddled with liver spots.

"VIP," he says with a smile.

I look down at my clothes – a tattered red shirt, a pair of mildewed shorts, sandals caked in jungle mud. I shrug and the Creole man laughs. I join the line of dignitaries and sneak into the Palace.

On Independence Day in 1975, citizens in the new Republic of Suriname enjoyed one of the highest standards of living in the Caribbean. The economy was robust, if small, buoyed by the soaring price of bauxite. Foreign reserves were growing steadily and foreign investment in Suriname's rich mineral resources was showing signs of an upsurge. The Dutch had pledged a healthy 3.5 billion guilders in development aid, with the goal of making Suriname economically self-sufficient in just ten years. The time seemed ripe to break from the motherland.

Today, Suriname is one of the poorest and least-developed nations in the region. The economy is in an inflationary tailspin, state corruption is rampant, and the recently minted SRD, or Surinamese dollar, is almost worthless. Approximately 70 per cent of the population lives below the poverty line and the country has suffered two *coups d'état*, nearly a decade of military dictatorship, and six years of civil war since breaking the bonds of *patria*. Independence, which was supposed to bring prosperity to the young republic, delivered instead hopelessness, violence, and tragedy.

Suriname chose independence in a reluctant, half-hearted way. Minister-President Henck Arron's announcement in February 1974, that Suriname would be sovereign by the end of 1975, came as a shock to many. No referendum was held and the vote in Parliament passed by only a narrow margin. The Dutch were eager to sever ties quickly and cleanly, especially in light of their calamitous dealings with Achmad Sukarno over Indonesian independence in the late 1940s. To avoid being labelled "neo-colonialists," and perhaps spurred on by what some observers termed "Calvinistic guilt feelings," the Dutch ignored all protests and cooperated fully with the Arron government.

Prior to 1975, politics in Suriname had been fairly stable. A balanced power-sharing structure had emerged between the main Creole and Hindustani parties. But with independence came a frantic grasp for power in which ethnic loyalties soon hardened into ethnic chauvinism. An alphabet soup of political parties vied for control. Graft and corruption became widespread. Parliament broke down into a *"circus stupido,"* wrote one observer, as members were thrown out by the police, the sitting government was accused of fascism and dictatorship, and proceedings were routinely shut down for months at a time.

Then another tension emerged – a battle between the established elites, or *oude rotten* (old rats), and the young political upstarts, or *ruziemakers* (troublemakers). When a *ruziemaker* tore up a copy of the Rules of Procedure and scattered it like confetti across the floor of the National Assembly, the newspaper *De Ware Tijd* suggested that "Suriname has probably reached a new low in parliamentary rowdyism."

35

A fearful unease began to spread throughout the new republic. The confidence and slim sense of nationalism that had accompanied independence was beginning to wane. In the "golden handshake" agreement with the Dutch, Surinamese citizens were permitted to emigrate to the Netherlands for free until 1980. More than one-third of the citizenry now chose to leave. The exodus grew in the violent decade that followed and continues to this day. In the thirty years since independence, Suriname's population has grown from 350,000 to only 420,000 while the number of expatriate Surinamese in the Netherlands has nearly tripled to over 320,000. In a sort of "colonization in reverse," to steal a line from the Jamaican poet Louise Bennett, these numbers are expected to equal each other by 2015. In a sad irony, the Netherlands is now the paradise the Surinamese seek.

In the run-up to elections in 1980, while the nation's attention was focused on the circus in the National Assembly, trouble was brewing in the barracks of the Surinamese National Army. Officers who had trained or served in the Netherlands were being paid much higher salaries than new recruits. Desperate soldiers were forced to become businessmen on the side just to survive. Demonstrations aimed at forming a military union were stamped out with police force and repeated arrests. The soldiers, already unclear about the military's role in the new republic, became increasingly disillusioned, their morale sinking by the day. Tensions escalated when three high-profile officers were arrested and tried for sedition. A violent uprising was gathering force.

On the morning of February 25, 1980, a small team of commandos led by Master Sergeant Desi Bouterse stormed the Memre Boekoe Barracks, killing one policeman and the on-duty lieutenant, and raided the store of weapons. Another group of rebels seized the

marine base, drove two patrol boats upriver to police headquarters on the Waterkant, and opened fire. The building was destroyed and the munitions bunker was seized. By midday, the police commissioner had surrendered and the fighting ceased.

At 7:30 p.m. the next day, the vice premier of Suriname and the minister of justice and police formally turned the government over to Desi Bouterse's military. All told, five men died during the coup. Suriname's experiment in democratic self-rule had come to a crashing end.

Inside the Palace gates an opulent garden party is gathering steam. A small military band of flutes, clarinets, and a booming tuba entertains the crowd, and beneath a huge pavilion visiting dignitaries drink faux-Prosecco and smoke Cuban cigars. Small groups of Creole women dressed in colonial-style skirts and head-wraps sit beneath spreading mango trees. Every now and then, a mango plummets to the ground and one of these women scream. The Peace Corps are here; the young Americans staying at the same guest house as me. So are volunteers from Canadian Crossroads. Rich businessmen and gorgeous celebrities mingle with government ministers and foreign ambassadors. Someone calls my name and I whirl around. Toni, a part-Amerindian I met in Raleighvallen, is standing behind me with a man I don't recognize.

Toni is an expert wilderness guide. I first met him as he was returning to Foengoe Island from a day hike with a group of tourists. His tour group was clearly frightened, and as I got close I could see why. In Toni's arms was a massive bushmaster, one of the most poisonous snakes in South America. He'd caught it with his bare hands.

"What the hell are you doing here?" asks Toni.

"Research."

"This is Bodi." Toni's friend gives a wan smile and reaches out his hand. He looks angry, like a tough brawler who's just lost a fight.

"What are you researching?"

"Bouterse."

Bodi sucks his teeth, gives me a strange look.

"I'm kidding," I say.

"Andrew's a writer," says Toni, laughing. "Everything's research to him."

"How'd you get in here?" Bodi asks.

"I walked."

"Relax, Bodi," says Toni. "It's a party."

Bodi smirks. "Every year, more Americans, *toch*."

"I'm Canadian," I say. "And I was kidding."

Bodi says nothing and walks away.

"Touchy guy."

"He used to be in the army," says Toni. "All of them are touchy."

"Hey, nice suit."

"Boy!" yells Toni, fingering the mud on my collar. "It's *Srefidensi*! Show some respect!"

I walk up the terrace to the Palace and wander through the mob of people marvelling at the elegant interiors, the white walls hung with priceless European paintings, the sculptures, the elaborate chandeliers. The tiled floors are covered in red carpet. Every few seconds a flashbulb goes off, and the onlookers spin around to see who has arrived.

The howl of a loudspeaker draws me onto the balcony overlooking the square. Thick plumes of green and red smoke drift across the lawn. The crowds are still here, but now everyone is looking to the sky. Far to the west, perhaps a mile above the city, four paratroopers drift lazily through the clouds. The voice on the loudspeaker gives a running commentary in Dutch, and as the parachutists get closer the crowd begins to cheer.

But something is terribly wrong. One of the jumpers is well below the others and falling fast. His parachute is open but he appears to be out of control. The colour commentator yells and the people scream, but there is nothing the trooper can do. He plummets into the skyline of the Old Town, likely somewhere near Stadszending, and hundreds of onlookers rush up Lim A Po Straat to see if they can help.

The next man lands safely in the square, to the delight of the audience. Then the loudspeaker squeals again and we all look up. The last two are faring badly. One of them is headed straight for the majestic bell tower of the Ministry of Finance building, originally built in 1836 to house City Hall. His comrade is far to the north and making a beeline for the steeple of the Peter and Paul Cathedral, the tallest timber-frame structure in the western hemisphere. The onlookers howl in a mixture of fear and excitement. The loudspeaker announces a sudden drop in wind and falls silent. Both paratroopers are swallowed by the city, and two more units of concerned citizens pour up Gravenstraat and along Mirandastraat to rescue them.

I wander back to the gardens, where the guests have gathered at the foot of the terrace with drinks in hand. Above them, a pair of waiters dressed in white hover around a mahogany podium, behind which stands a phalanx of immaculately dressed women and men. We wait. No one speaks. Another mango plummets. And then the president appears.

Desi Bouterse is the most controversial figure in Suriname's young history, a history that still reverberates today, I've come to realize. Blessed with charisma and a penchant for dramatic, emotional speeches, he is a complex, unpredictable, and often brutal character. A former soldier and sports instructor who was also an accomplished long-distance runner, a former *ruziemaker* who once supplemented his military salary by farming chickens and pigs, Bouterse remains a divisive character some twenty-five years after he burst onto the national scene.

Most Surinamese greeted Bouterse's coup with euphoria or quiet passivity. The daily *Vrije Stem* ran the headline "FINALLY!" and editorialized that most Surinamese supported the coup because "they feel that we have totally run aground in the old politics." The coup was viewed as an *ingreep* at first, an intervention aimed at removing the *oude rotten* from power and instituting a new "people's democracy." Bouterse and his army were the heroes of the new republic.

The early days of military rule were tinged with a charming innocence. Bouterse became the leader of the ruling National Military Council (NMR) and instituted a civilian government led by the physician Henk Chin A Sen. Both pledged to uphold the tenets of the Surinamese Constitution of 1975 and committees of citizens, or *volkscomites*, were established throughout the countryside. To demonstrate the new leaders' openness, a suggestion box was posted at the gates of the Memre Boekoe Barracks. A musical cabaret about the coup – *Ba Uzi*, or Brother Uzi – was allowed to run for two weeks in Paramaribo. One song captured the exhilaration and unease of the times:

A first coup is like a first love . . .
Love is true – love is blind
Everything shines in a golden light
Everything looks bright!

But it wasn't long before cracks began appearing in Bouterse's *ingreep*. Rumours of a planned "coup-within-a-coup" surfaced, and Bouterse had several of his supposed supporters arrested and sentenced. He then dissolved Parliament and his Advisory Council and reinstituted a curfew. Rivalry between the army and the NMR worsened as Bouterse stepped down from the council to focus on his role as army commander. A manifesto began to circulate that promised to "thoroughly raze the false parliamentary democracy of neo-colonialism." Bouterse shelved the constitutional reform process, postponed promised elections, and formed a new leadership council, the Revolutionary Front, inviting delegates from socialist parties across South and Central America and the Caribbean to the inauguration. His government survived a series of coup attempts, and then he ousted Chin A Sen from the presidency, much to the consternation of the Dutch.

Bouterse began making increasingly radical statements that echoed the revolutionary fervour of his counterparts in Cuba, Nicaragua, and Grenada. "We have stooped long enough beneath the yoke of capitalism," he said. "We're all striving for a socialist society in which there is work for us all, where social justice reigns, and in which there is no longer any poverty, exploitation, racism, or oppression." All noble causes, to be sure, but Bouterse was not a leftist by anyone's standards – his guiding ideology was the retention of power at all costs. As military rule dragged on, the citizens began to fear for the future and a public backlash began.

Labour unions went on strike. Student organizations demonstrated in the streets. The Council of Christian Churches, the Bar Association, and the Association of Businessmen all voiced their protest. Bouterse's more leftist friends abandoned him as a series of economic scandals rocked the military. Most significantly, the press began criticizing the regime. No matter the intimidation tactics used by the military, there seemed no way to keep the people out of the streets. The people of Suriname had finally rallied around a common cause. Tragically, their actions came too late.

In late October 1982, Grenada's revolutionary leader, Maurice Bishop, paid a visit to Bouterse in Paramaribo. The New Jewel Marxist cautioned the young army commander: "The Surinamese revolution is too friendly. Reactionary forces are too strong. You have to eliminate those who are not with you; otherwise they will eliminate you." Bishop's welcome reception at the Palace had to be lit by candlelight, as recent strikes had knocked out the city's electrical grid, and Bouterse took his advice to heart. The violent prophecies in the cabaret, *Ba Uzi*, in which power-drunk liberators eventually turn their guns on the populace, were about to come true.

On the night of December 7, 1982, the headquarters of two daily newspapers and two major radio stations were firebombed by soldiers armed with grenades and bazookas. Bouterse's men swept through the city armed with a list of "enemies" and arrested sixteen men – prominent military officers, lawyers, journalists, radicals, and academics. The detainees were brought to the Zeelandia prison, where they were interrogated and tortured. The next day, fifteen mutilated and gunshot-riddled corpses were dropped off at the Academic Hospital morgue.

The events at Fort Zeelandia on December 7, 1982, are still shrouded in mystery. The official line, delivered by Bouterse on television the next day, was that the prisoners had been shot during an escape attempt. But this story conflicts with the fact that many of the bodies had clearly been tortured and shot from the front, as if by firing squad. Soon countless versions of the truth were circulating – that Bouterse had led the interrogations personally, that he had pulled the trigger himself, that he hadn't even been at Zeelandia that night, that he was fast asleep at the home of his friend Lie Pauw Sam when news of the murders reached him. One prisoner, the firebrand trade unionist Fred Derby, somehow survived the shootings (Bouterse claims he arrived at Zeelandia just in time to spare him). For the rest of his life, during which he became a much-loved politician, Derby famously refused to speak of the events of that night. He died of a heart attack in 2001, taking the truth to his grave.

The December Murders, as they came to be known, marked another turning point. The public was stunned and traumatized, unable to believe that their small, quiet country could erupt into such cold-blooded violence. Domestic support for Bouterse disintegrated and the Dutch suspended their "golden handshake" development payments indefinitely. The international community condemned the killings amid fears Suriname was sliding toward Communism – the last thing anyone wanted in 1982 was another Cold War battleground. All foreign aid was cut off. Five days after the killings, thousands of brave mourners turned up for the funerals, after which the National Women's Council led thousands of women in a march on Independence Square. As they passed the Dutch embassy, a spontaneous cry of "Help us, help us!" rose up from the crowds.

No one has ever been held responsible for the murders. The event is arguably the key psychological trauma of Suriname's modern history, an incident that perfectly encapsulates the powerlessness, hopelessness, and fear that pervade the country, in some respects, to this day. In the aftermath of the killings, amid so much anger and uncertainty, only one thing was clear: on December 7, 1982, the young Republic of Suriname lost her innocence.

President Ronald Venetiaan steps to the podium. Surrounded by his family, members of his cabinet and hard-faced government ministers, he gives a short speech, half in Dutch and half in Sranantongo. He is interrupted a number of times by raucous applause. Then the waiters deliver flutes of champagne to everyone on the terrace. Toni nudges me.

"There's Suriname, boy," he whispers, his eyes on the spectacle above us. "The rulers drink champagne, the people drink fake sparkling wine."

Venetiaan walks to the front of the terrace and leads the crowd in the Surinamese national anthem. After the wearisome last lines, during which we pledge to fight for Suriname until our death, another song begins. Now everyone sings with more force and passion. An enormous smile peels across the president's face. This is Suriname's unofficial national anthem – Mi Kondre Tru, mi lobi yu – the theme from the popular musical of the same name based on the tragedies of Suriname's revolution. My true country, I love you.

As the song dies down, I see a face I recognize in the crowd. I ask Toni.

"You got it, boy," he says. "That's him." Ronnie Brunswijk, member of parliament, king of the Jungle Commando.

Desi Bouterse – as all worthy protagonists should – has a nemesis who is just as charismatic as he. His name is Ronnie Brunswijk.

In the years following the December Murders, Suriname's small economy continued to crumble, the international community strengthened its boycott, and the public grew ever more fearful. Bouterse proved himself a minor political genius as he skilfully negotiated continual infighting within his revolutionary ranks, a paralyzing bauxite strike, drastic food shortages resulting in bread lines, rising unemployment despite a tidal wave of emigration to the Netherlands, the growth of Suriname's famed black markets, and constant accusations from the Netherlands and the U.S. that his military was providing a lucrative cover for Colombian drug cartels smuggling cocaine to Europe (this was true). Bouterse continued to give stirring speeches, promising a new constitution and a rapid return to democracy, and he even consulted with the *oude rotten* he had ousted in 1980, Henck Arron and Jaggernath Lachmon, on the formation of his cabinet. At the same time, he cagily refused to commit his regime to any one ideology, playing what the political writer Edward M. Dew calls a "constitutional shell game" in which his opponents were left guessing. He flew to an international conference with his confidante, Maurice Bishop of Grenada, on Fidel Castro's plane, paid a visit to Muammar al-Qadhafi in Libya and repeatedly accused the CIA of plotting his overthrow (also likely true). But he also ousted the Cuban ambassador when U.S. troops invaded Grenada. By 1986, the Surinamese guilder had sunk from 1.8 against the dollar to more than 10, but still Bouterse appeared resolute. He announced that the seven "lean" years were over, and the seven "fat" ones were about to begin.

He was badly mistaken.

Deep in the jungles southeast of Paramaribo, a small guerrilla force of Ndyuka Maroons calling themselves the Surinamese Liberation Army (SLA) launched a violent campaign to oust Bouterse and his military cronies from power. They were led by Ronnie Brunswijk, an enormously muscled young Maroon who had recently been fired from Bouterse's security detail over a pay dispute. In the early days of his rule, Bouterse had pledged his allegiance to the country's most marginalized people, and no Surinamese citizens were more disenfranchised than the Maroons of the interior. Bouterse provided many with employment either as soldiers or bodyguards. But when Brunswijk's men began their assaults, including an attack on a police post in Albina, Bouterse retaliated with the full force of his military, sending troops into a number of Ndyuka villages in search of "the terrorists." A civil war had officially begun between the Maroons and the Surinamese Army. The SLA began calling themselves the Jungle Commando (JC), and Commander Brunswijk soon found himself cast as a sort of Robin Hood of the rainforest. By the end of 1986, three hundred Maroons had joined the cause.

The war raged, off and on, for the next six years. The violence occurred far from Paramaribo, in the east and southeast. If the December Murders are the most significant trauma of Bouterse's regime for those in the city, the Hinterland War, as it came to be known, continues to haunt the country's interior. The resistance fighters became, in the eyes of many, modern incarnations of the courageous African slaves who threw off the shackles of slavery and fought against the Dutch military in the seventeenth and eighteenth centuries. Relations between the Maroons and the black Creoles in the city had always been strained – Maroons hold that the Creoles are the descendants of weak-willed slaves who were not brave enough to escape the plantations, while the corresponding Creole axiom is that the Maroons are backward, half-naked primitives. Now Bouterse and Brunswijk became the opposing faces of these racist beliefs.

During the war, nearly every Maroon village in eastern Suriname was razed either by soldiers or by rockets fired from Alouette helicopters. Much of the town of Albina was burned to the ground as the Marowijne River Region erupted in violence. More than ten thousand Maroons fled across the river to French Guiana, where the government set up refugee camps. Whispers of supernatural forces working on behalf of the JC – the *winti* (spirits) and *obiamans* (shamans) of Maroon culture – spread through the city. There were also rumours of the army importing thugs and rockets from Libya to help put down the insurrection. The JC made a special target of Suralco, the Surinamese bauxite company, as well as outlying military posts, while Bouterse attacked Ndyuka and Saramaka Maroons, Brunswijk's main followers. In the village of Moiwana, thirty-nine defenceless Ndyuka were slaughtered by the military.

In 1992, the fighting finally came to an end with the signing of the Peace Accord. The following year, Bouterse stepped down from his role as army commander and sought more legitimate power as the figurehead of Jules Wijdenbosch's National Democratic Party (NDP). Parliament was reestablished under President Ronald Venetiaan and his New Front coalition, and with the help of development funds from the Netherlands, the International Monetary Fund, and the World Bank, inflation began to ease (the guilder was now worth a pitiful 27 against the dollar). But in the elections of 1996, Venetiaan's New Front lost to the NDP, which set about replacing the previous government's structural adjustment programs with a series of loose fiscal policies. Wijdenbosch's actions, with Bouterse pulling strings behind the scenes, ushered in a new era of rampant government corruption, debilitating inflation, soaring exchange rates, and economic stagnation. Though Venetiaan won back the presidency in 2000, Suriname's economy has been in steady decline ever since.

In an absurd development typical of strongman politics, both Desi Bouterse and Ronnie Brunswijk now serve as elected members of Suriname's current parliament. Even more ridiculous is that Bouterse continues to pursue the presidency even though he is effectively unable to leave the country. In 1999, he was convicted *in absentia* for money laundering and cocaine smuggling and sentenced to eleven years in a Dutch prison. The Netherlands maintains an international warrant for his arrest.

An hour later, Toni and I are standing in a long line of well-wishers outside the presidential ballroom. The noonday sun has begun its burn, and the crowds in the square have long since retreated to the shade.

"You ever heard of *okopipi*?" asks Toni.

"Huh?"

"The blue frog?"

"Blue frog? No."

Toni frowns. "Nobody knows how to sell this fucking country. *Okopipi*. The tiny blue frog that only lives in Suriname."

"Never heard of it."

A large Creole women dressed entirely in green approaches. She tries to sell me her chapbook of poetry. When I politely decline she smiles at me through the lime leaf in her mouth.

"You can only find it in one little mountain range," says Toni.

"This might be the only country where you can buy poetry while waiting to shake the president's hand."

"Boy, listen! The blue frog. It only lives in the south, in Trio territory near Brazil."

"Yeah?"

"Yeah! Nowhere else!"

"Have you ever seen it?"

Toni frowns. "No."

"Why not?"

"I'm not allowed down there." Toni pretends to shoot a rifle at me.

"Then how do you know it exists?"

Toni shoves me in the chest. "I guess I don't. But if you want to know about Suriname, boy, you've gotta see it. It's only this big." He holds his thumb and index finger an inch apart. "*Okopipi*. The blue jewel of the jungle!"

I like the sound of this.

"So how do I get down there," I ask.

"You need to get a permit."

"Where?"

Toni shrugs, does the rifle thing again. "No fucking clue."

Finally, it's my turn to meet President Venetiaan. He smiles, shakes my hand, and I lean in to speak.

"You have a beautiful country," I say.

"I am happy you think so," he says.

"Your forests are the finest in the world," I say.

"Thank you," he says. "I agree."

Venetiaan then introduces me to his wife, a striking woman in immaculate white robes embroidered in green and gold, a string of pearls around her neck. She takes one look at my soiled clothes, muddy sandals, and the band-aids on my toes and bursts into laughter.

After a long night of street parties, I spend the morning nursing my hangover in an internet café, researching *okopipi*.

In 1968, a young Dutch herpetologist named Marinus Hoogmoed discovered this remarkable species of frog while exploring the valleys of a modest set of mountains in Suriname's remote Sipaliwini Nature Reserve. He called it *Dendrobates azureus* after its stunning azure-blue skin. The Trio Indians, in whose territory Hoogmoed was travelling and who had known of the frog's existence for centuries, already had a name for it. They called it *okopipi*.

Okopipi is revered by the Trio for its rarity, its iridescent blue skin, and the extraordinary strength of its poison. Each tiny frog contains an average of two hundred milligrams of toxin; two milligrams on the tip of an arrow is enough to kill a man. The Trio used to catch *okopipi* and sell them on the exotic animal market, but laws have recently been passed banning their trade. Today, *okopipi* is one of the most vulnerable frog species in the tropics of the New World, a worrying claim to fame considering that frog populations – long considered the canaries in the coal mine of ecosystem health – are in severe decline throughout the world. According to the World Conservation Union, at least one-third of all known amphibian species on earth are on the verge of extinction. The total number of *okopipi* is thought to be no more than three hundred, and they can be found only in the mountain valleys on the Brazilian border, where Hoogmoed found them.

The more I read, the more fascinated I become. For an ex-biologist obsessed with locating Suriname's quintessential soul, *okopipi* seems the perfect quarry – elusive, endangered, desperately rare. The blue jewel of the jungle might be the perfect metaphor for a country so ecologically, economically, and politically fragile, the delicate spirit of the Last Eden.

I decide to travel to Brokopondo tomorrow and spend the next month exploring the lake. But I will also begin planning an expedition to find this little frog.

In the late afternoon, someone knocks on my door at Stadszending. Through the window I see Toni's brooding friend Bodi, a motorbike helmet under his arm.

"Open," he says through the glass.

"Why?"

"For talk."

I think quickly. Outside. Outside should be OK. How the hell did he find me? I unlock the door, open it a crack.

"How'd you find me?" I ask.

"Not hard," he says.

"Give me a second." I shut the door, lock it again, grab my smokes, take a breath. Be cool. Outside is fine. "One second!" Where are those beers? Be cool! I find two warm bottles of Parbo under my bed.

I open the door, hand Bodi a beer, try to act as if nothing is strange about a military man tracking me down at my guesthouse. This proves difficult. We sit on the terrace outside my room. I light a cigarette, stifle the fear in my gut. Bodi sits calmly, his arms folded across his chest.

"You are journalist?" he asks, helping himself to a Morello cigarette.

"Not really."

"What, then?"

"A writer."

Bodi takes a long drag. "No difference."

"OK."

"You talk about Bouterse."

"I was kidding."

"You're lying."

I stare at him for a moment.

"Did I offend you?"

"You're lying." Bodi shuts his eyes as he exhales.

"OK," I say. "I'm interested in Bouterse."

"Why?"

"Because he's interesting."

Bodi shakes his head. "It is dangerous, *toch*." The Dutch expression *toch* is short for *Yu sabi, toch?* It translates roughly as, You know what I'm saying?

"What's dangerous?"

51

"People see you talking to me, they talk."

I watch Bodi while he drinks. He is at least forty but has a young, chiselled face that looks vaguely Asian but also African. His body is a blend of thick muscles and skinny limbs, a strange combination of bundled power and wiry toughness.

"Are you Javanese?" I ask, changing the subject.

"I am Javanese, Creole, Brazilian, Chinese, Amerindian, and Jewish," he says with a smirk. "Perfect Suriname man."

"And you were in the army?"

"Who told you that?"

"Nobody." I try to smile. "I just thought maybe –"

"Who fucking told you that?"

I swallow. "Toni."

Bodi sucks his teeth. "Toni knows shit, *toch*."

"You're right. You're absolutely right. Toni knows shit." I remember the cigarette in my hand, take a drag. A stub of ash drops on my knee. Bodi says nothing, just sucks on his Morello as if sucking the air straight from my lungs.

"So, you weren't in the army," I say.

"I was."

"Oh. So, when did you get out?"

"I'm not out."

"You're still in?"

"I am in and I am out."

I look at Bodi. He frowns, rolls his eyes.

"They call me," he says, pointing to the mobile blinking on his hip. "When they have mission."

Bodi's hands are too big for his arms and are pocked with scars, like those a child might get from a bad case of chicken pox. Bodi, though, does not strike me as the kind of man who would ever have scratched himself raw. On the contrary, he seems like someone with a very high regard for self-discipline. He is also clearly comfortable with silence.

"What do you do in the army?" I ask after a while.

"Lifeguard."

"Huh?"

"Lifeguard, *toch*."

"You mean bodyguard?"

"Bodyguard. Yes."

"For who?"

"Who do you think?"

It takes me a moment. "For Bouterse?"

Bodi nods. Then he sees the look on my face.

"Don't worry."

"Easy for you to say."

"No," he says, shaking his head. "It's not good job."

I stare at Bodi for a moment. "Then why don't you quit?"

"If I quit, I have to move to Holland."

"Why?"

He doesn't answer.

"You don't like Holland?"

"This is my home."

"Of course."

"Suriname is rich country. We have gold, bauxite, diamonds, oil. So little people. So why aren't we all fucking rich?"

"*Me no sabi*."

"Government." Bodi frowns, rubs his index finger and thumb together in the air. "Fucking government."

"Same everywhere," I say.

Bodi lights another smoke. "What you want to know about Bouterse?"

"What it's like working with enemies."

"Everyone in Suriname is enemies," he says. "Everyone in Suriname is friends."

Silence.

"Do you know the truth?" I ask, stubbing my smoke beneath my chair to hide my nerves.

"About what?"

"1982."

I look up slowly. Bodi is staring at me. His jaw ripples with anger. "I don't talk those things."

"Sorry."

"I have kids, *toch*. Family." Bodi's phone vibrates. He checks it quickly.

"I'm sorry," I say again.

"My wife call. I go now." He stubs out his cigarette. "Just I try to forget."

Bodi makes no effort to leave. He sits there, his last words hanging in the air like an invitation. My mind races. A thousand questions tussle for room in my head.

"Forget what?" I ask.

"People killed," says Bodi quietly.

"Which people?"

"Doesn't matter."

"People you killed?"

Bodi snaps out of his daze and pushes himself up from his seat. "Thank you for the beer."

"No problem."

"Give me your mobile. Maybe I call you sometime."

I take his phone, type my number into it.

"Maybe we play football," he says. "Something."

"I'm headed south in two days."

"*Pe yu e go?*"

"Brokopondo Lake."

"How long?"

"Maybe a month."

"Why?"

"I'm not sure."

Bodi takes two Morellos for the road, walks down to the courtyard, and picks up his motorbike from the dirt.

"Andrew!" he yells up from the yard.

"What?"

"What you want to know?"

I lean over the railing. "Everything," I say.

"You can't."

"Why not?"

"Because," says Bodi, kicking his bike to life. "Nobody knows
everything."

The next morning, I walk down to the Waterkant, grab a beer, and
head for Fort Zeelandia. I pay the entrance fee and wander through
the museum – a colonial-era pharmacy and doctor's office, an
exhibit of Amerindian artifacts, a selection of photographs from a
nineteenth-century leper colony, a dank dungeon filled with the
vicious hooks and knives of torture. Then I walk out into the courtyard and up a flight of stairs to the balcony that overlooks the river.
From here I can see the District of Commewijne, the nameless
bridge over the Suriname River, the tugboats and sailboats riding
the ebb tide. This is the perfect place for a military base; from this
bend in the river, you can see up and down the Suriname for miles.

I walk to the southeast corner of the fort. Down below sits the
statue of Queen Wilhelmina of the Netherlands, surrounded by
ancient canons lying heavy in the sand. Then I inspect the balcony
wall and find the bullet holes. I wouldn't have noticed them if I
hadn't been looking.

I try to imagine where the victims were standing, which way they were facing as they tried to escape, or as the firing squad took aim. I wonder if they were looking upriver toward the jungle or downriver toward the sea. At my feet sit three red pylons, the Surinamese equivalent of police tape. The bullets removed from these holes are the only remaining physical evidence of the December Murders.

If the December Murders perfectly encapsulate Suriname's loss of innocence as a young nation, Bodi might be the perfect human embodiment of this trauma, a man clearly unhappy with his lot but nonetheless resigned to his role. I suspect he knows the truth about those killings. I suspect a part of him – the less disciplined part – would like to share his thoughts with someone he could trust.

I leave the fort and head north on Kleine Waterstraat. At the STINASU offices I speak with a conservation official, an older Creole man whose official title is impossible to discern. I tell him I'd like a permit to visit the Sipaliwini Nature Reserve, the home of *okopipi*.

"You are scientist?" he asks.

"No," I say. "Not any more."

"Sorry," he says, shaking his head. "Permits are only for scientists."

2

THE RED ROAD

At seven o'clock on a weekday morning, Saramaccastraat is alive with nervous energy. Huckster shopkeepers stand in front of their tiny stores with fierce gazes, toothpicks poking from the corners of their mouths. Street peddlers, many of them old Maroon women with colourful hair-wraps, crouch on the sidewalk behind blankets piled with bunches of bananas and pyramids of oranges, torn pieces of cardboard advertising their prices. A giant man in a bright green T-shirt sells burned CDs from a plank of wood balanced on his belly. The Central Market has just opened and the smell of fish and the sweat of nocturnal fishermen wafts from its alleyways. Gangs of young drug dealers and gold miners and aspiring reggae singers swarm the northwest corner of Dr. Sophie Redmondstraat in a cloud of marijuana and macho bravado. A dandy in a baby blue suit and grey fedora leans on his golden cane and harasses the women who walk past, regardless of age or figure or stage of pregnancy. An army of white cargo vans, the ubiquitous *wagis* of Suriname, now stained red, lines both sides of the street, their owners scrubbing the bauxite dust off them in preparation for another coating of bauxite dust. And hovering near these vehicles, groups of southbound Maroons gather in the meagre shade of nearby shop-fronts, jerry-rigged backpacks and double-knotted plastic bags at their feet, waiting for the journey home to begin.

For the Maroons of Brokopondo and Saramaka, there is only one way to get to the city and there is only one way to return home. All of these journeys either begin or end here, on Saramaccastraat, the Grand Central Station of Suriname.

I find Kevin buying groceries in a supermarket filled with chocolate bars and booze. Kevin is a Peace Corps volunteer, a PCV, one of approximately forty young Americans working in Suriname. We first met a month ago at Stadszending, and he gave me a standing invitation to visit him in the Saramaka Maroon village of Balingsoela, his home for the next two years. Balingsoela is less than ten kilometres from the lake, the perfect base from which to explore Brokopondo.

Kevin is headed back to the interior after an extended medical leave in Parbo – two weeks ago, he became the latest PCV to contract dengue fever. His was a mild case – he suffered none of the horrific symptoms of "breakbone fever," the "bonecrusher disease." Suriname's interior is currently being swept by a dengue epidemic for which there is no vaccine.

Kevin leads me to his *wagi*. The driver, a boy no older than sixteen, happily shakes my hand and takes my backpack. He unties a piece of twine at the rear of the van and the back door flies open. He shoves my pack on top of a pile of bags and propane tanks and then pulls the door shut again, holding it in place with his foot as he reties the twine.

"*Oten wi gwe?*" Kevin asks the boy. When do we go?

"*Dalek,*" says the boy.

Stolen from the Dutch, the Sranantongo word *dalek* translates as "soon" or "in a bit." Its real meaning, though, is closer to "it'll happen when it happens." *Dalek* is every Surinamese's favourite response to any question concerning time. It can be a frustrating answer for a visitor to receive, especially when making appointments or travel plans, but there is a positive side to *dalek* as well. The word comes with a guarantee that whatever you are asking about will eventually happen, just not yet, and can be a soothing revelation to travellers in unknown countries.

Kevin smiles and turns to me. "You know about *dalek*?" he asks, as he pops the cap on a Parbo.

"Yeah," I say, taking the beer and breaking my fast.

Kevin sits down on the curb and fumbles for his Morellos. "I'm going to get *dalek* tattooed on my fucking chest."

Ninety minutes later, half of the *wagis* have already left for the south, but not ours. Finally, a group of Maroons emerges from a nearby shop. Without a word, the driver jumps into the truck. Kevin and I climb in the side door and settle into the back seat. The *wagi* is designed to seat nine, but by the time the doors are finally closed there are fourteen of us squeezed in among hundreds of pounds of gear.

We pull out into the street, bottom heavy and flesh filled, and head south past flashing casino lights. We cross a small bridge, where an old Rastafarian man balances on a rocky ledge and serenades the passing traffic through an imaginary microphone. Although it feels like we're about to leave the city and begin our journey, this is an illusion. We stop at Popeye's Chicken and Biscuits so the old woman in the front seat can get her fix. We pull in at a rural grocery store, where two kids hop out and return with bags of white bread. At a construction depot that doubles as an abattoir, the driver haggles with a Hindustani man dressed in a blue, bloodstained smock; then the two of them shuffle five thirty-pound bags of cement over to the *wagi*. Just as I wonder where these bags are going to fit, the driver finds just enough space for them beneath my feet.

The supply run continues. We stop for six watermelons at a road-side stand. We slow for mangoes but decide against them, to Kevin's dismay. Finally, an hour after leaving Saramaccastraat, we pull in at an old tavern for a rest. The men swarm the caged-in bar while the women perspire on white plastic garden chairs. We have covered a total of ten miles, we haven't even left the city yet, and everyone is hungry and exhausted.

Soon we're back on the road. The *wagi* rattles over potholes on the Martin Luther King Highway as we race through the savannah belt south of the city. The cab echoes with the rustling of plastic bags as we devour trays of *bami kip*, the Hindustani staple of chicken and fried noodles. Soon we reach a rare intersection, where a billboard shows a tiny, defenceless car at the moment of annihilation by an enormous, barrelling dump truck. We have reached Paranam, home to Suriname's famed bauxite mines, Suralco's flagship alumina refinery, and to its now-closed aluminum smelter.

Bauxite, the raw material of aluminum, is the backbone of Suriname's economy. The coppery soil accounts for more than 70 per cent of the country's export earnings and the deposits here are among the richest in the world. During the Second World War, Suriname was the main bauxite supplier for the aluminum-hungry U.S. Air Force – the country's international airport was originally built by the Americans. Without its lucrative red earth, Suriname would hardly have a legal export economy to speak of – the nation's other top earners are, unofficially, casino revenues and drug trafficking. Between one- and two-thirds of all narcotics in the Netherlands are thought to originate from Suriname's jungle airstrips, rumoured to be one of the main trans-shipment points in the Americas for Colombian drug cartels.

In the early 1960s, Suralco, a subsidiary of the American aluminum giant Alcoa, embarked on the most ambitious development project in Surinamese history – the construction of Suriname's first alumina refinery and aluminum smelter. Bauxite is worth more on the export market if it is first refined into alumina, and even more if the alumina is then smelted into aluminum. This would strengthen the nation's otherwise struggling, import-based economy. There was only one catch: turning alumina into aluminum requires an enormous amount of electricity. Where would Suralco find the power?

Sixty kilometres south of Paranam, the massive Suriname River snakes north to the Caribbean Sea from twin sources deep in the Guiana Highlands. The Suriname is the country's largest and most celebrated waterway. On its northern shores, coffee and sugar plantations first sprang up in the seventeenth century; on its southern shores the first rebel slaves made their homes; and it was near its mouth where the settlement of Paramaribo was built by the English and expanded by the Dutch. The Suriname River has always been the nation's economic and spiritual heart.

In a twisted sort of way, Suralco's vision for the river was in keeping with its coloured past. On February 1, 1964, after more than two years of construction, Suralco sealed its hydroelectric dam across the Suriname River at a tiny port called Afobaka. Over the next two years, more than fifteen hundred square kilometres of pristine rainforest were drowned beneath upwards of thirty metres of water. Forty-three Saramaka and Ndyuka villages were destroyed and more than six thousand Maroons became instant refugees. By the time the waters reached their full depth, the aluminum smelter at Paranam had more electricity than it could use and Suralco had created one of the largest man-made lakes on earth. The reservoir is officially called Prof. Dr. Ir. W.J. van Blommenstein Meer, after the Javanese-Dutch hydrological engineer. To most, though, the reservoir is known simply as Brokopondo Lake.

Up ahead, Suralco's alumina refinery looms like a scene from a post-apocalyptic novel. Giant smokestacks spew endless plumes of black smoke into the air. Warehouses that would dwarf an airplane hangar sit heavy in the foreground and sinister light-posts tower above the scene. An office building sits in the middle of this grey landscape like a huge, all-seeing eye, its windows lit by an infernal light that I've heard can be seen from jungle hilltops 150 kilometres away.

We turn off the concrete freeway and circle the refinery on a wide, dirt track, its surface covered in waste bauxite. The road turns south and our boy-driver punches the accelerator. The *wagi* rattles mercilessly as we gather speed. We have reached the infamous Afobaka Highway, Suriname's Red Road.

Our tires thunder as they slam into the ruts and crevasses that blight the road, each crash like a wrecking ball from beneath. Accompanying these blasts is the sickening sound of hundreds of pieces of metal shearing against one another. We all lean to the right as we swerve to avoid a particularly massive pothole. We swerve back to the left to avoid an oncoming tractor-trailer loaded with freshly cut trees. The road is a vehicular free-for-all with only one rule: the bigger the vehicle, the less likely it is to get out of your way.

For the first half-hour no one speaks, we just listen, Kevin and I gripping the seats in front, the Maroons accustomed to the experience but captivated nonetheless. No matter how often they've made this journey, the brutality of the Red Road requires a steeling of the body and mind.

Enormous DAF dump trucks race past, swamping our *wagi* in clouds of red dust. The pounding worsens. I imagine nuts and bolts shaking free from the chassis, screws loosening, the frame of our van blowing apart. I imagine fourteen cartoon people balanced on two axles and a plank of metal screaming down this jungle road, our driver gripping the wheel as if it were a live animal, a cigarette dangling from his bottom lip.

"The faster we go, the sooner it's over," yells Kevin, just as we hit a particularly rough patch and my head crashes into the roof. I think of V.S. Naipaul's trip down the Red Road in *The Middle Passage*. "We stared resentfully at the road," he writes. "We didn't see the forest. We saw only red dust."

Slowly, we work ourselves into the most comfortable position possible: backs straight, hands pressed down into the seats, shoulders wedged against our neighbours. Outside, small shrubs line both sides of the road. On the left, these bushes quickly give way to thick jungle, the vegetation on its leading edge stained pink with dust. To the right, the jungle sits back to make way for the enormous electricity pylons that run the length of the highway and bring power from the Afobaka Dam up to the Suralco plant. They stand like sentries watching over the road, a constant, haunting reminder of why this highway was built in the first place. Birds of prey circle above them, riding the thermals.

Up ahead, a lone man on the side of the road tries to wave us down. Our driver races past him but then thinks twice and slams on the brakes. A roar of disapproval rises from the cab – we were just hitting our stride, just learning or relearning how to sit this journey through. The worst thing about stopping on the Afobaka Highway is knowing that you have to begin again.

The hitchhiker runs up to the driver's open window and haggles with the boy. As they argue, the girls in front stare at the man and suck their teeth angrily. Just then, a DAF truck thunders past and a dark cloud of bauxite dust funnels into the *wagi* through the open window. If I were this hitchhiker, I would reconsider climbing into this cab.

Finally, the two men agree to terms and the side door slides open. Now we have our first glimpse of the man who made us stop. He is young, perhaps twenty years old, a shirtless Maroon with a slim, muscled upper body. Six-inch dreads stand straight up from his head and he wears nothing but a threadbare pair of Petrol jeans. He carries a small plastic bag, a box of toothpaste in the bottom, and as

63

he squeezes himself into the mass of bodies he smiles and gives a wan thumbs-up. He is a desperate sight, windblown and dustblown and obviously relieved to have found a ride. The door slams shut, our driver hits the gas and I taste the bauxite on my lips. Outside, a young boy pushes a wheelbarrow, his sister fast asleep in the hold.

A half-hour later we slow again. The driver leans out his window, inspects the front left wheel and pulls over. Everyone piles out, the women retiring to the shade of a nearby tree, the men relieving themselves in the roadside brush. The driver unties the back door and the pile of backpacks, grocery bags, propane tanks, and flats of cement tumbles to the red earth. He digs beneath the rear seat and retrieves a ridiculously small spare tire. The women mock him ruthlessly as he carries the tire with one hand to the front of the van. While the men take turns jumping up and down on the wheel-wrench to loosen the bolts, Kevin and I light a smoke and stand beneath a short, squat tree. Kevin tells me the same tire went flat eight days ago. They didn't have a spare then so the owner patched it by melting an old piece of tire and holding it in place until it stuck.

Kevin walks up the road while I check the reception on my cell phone. We are in the middle of the jungle, so no dice. One of the passengers approaches, a lovely Saramaka girl in a tight yellow skirt whom I hadn't noticed in the *wagi*. On her head, a black hair-wrap. On her face, disappointment.

"You number," she says.

"Sa?"

"You phone number," she says slowly.

I try to remember my cell number. I give it to her and she enters it into her phone.

"You name."

"Andrew."

She grimaces, raises the phone to her ear, and walks back to her friends.

The girls watch me. I'm not sure why. One of them wears a T-shirt that reads "Speed Limit 150 kph."

Out on the road, the hitchhiker attempts to abandon ship. He flails his arms, his bag of toothpaste whipping back and forth as he desperately tries to flag down one of the passing trucks. None of them stop and after a few minutes he returns to our sad little gathering at the side of the road.

Now the girl in yellow turns around and motions for me to raise my phone in the air. I do this and suddenly it rings.

I press TALK. "*Fawaka?*" I say. How are you? The girls collapse laughing.

"*Mi de,*" says the woman. "*Fawaka nanga yu?*"

"*Rostu.*" I'm good.

"*Mi Vivian,*" she says.

"*Mi Andrew,*" I say.

"*Mi sabi*," she says and hangs up.

Finally, the busted tire comes loose. The driver wheels it to the back of the *wagi* as the others bolt the new one into place. Beers are chugged, cigarettes are flicked. We pile our gear into the back again and climb in. With the absurd new tire, the *wagi* now leans precariously to the east.

For the next hour we careen down the highway, stopping at occasional villages like Koini Kondre, where Vivian and her friends pile out. Somehow, amid the bone-jarring thuds, the passengers find a way to sleep. They lean against each other, a tangle of unconscious limbs, heads, and torsos – struggling together, like any extended family would, to make the uncomfortable as comfortable as possible. Even the hitchhiker has found a place to rest his head.

For the Maroons of Brokopondo and of Saramaka farther south, the Afobaka Highway is the only way to reach medical help when traditional cures have failed, the only way to get to market when they have greens to sell, the only connection to the outside world should they need or want one. Saramaka men ply this route in search of employment on the coast. School-age Maroons whose families can afford tuition make the trip to Paramaribo every few months. Others, usually the elderly, travel the road perhaps once a year, to visit family or attend a funeral or deal with some citizenship annoyance, but only if they can afford the fare. The Afobaka is a costly highway in every sense: physically painful, financially dear, psychologically daunting, even emotionally jarring. Built to service the bauxite industry, the Red Road's role as the gateway to the interior seems almost an afterthought.

We turn off the main road and follow a slim, winding path into the jungle. Now our driver goes slowly and the forest closes in. Branches scrape over the roof and plunge into open windows. The *wagi* struggles until the driver remembers to shift down. Then we come to a clearing in the woods, a village of thatched huts and pre-fab buildings laid out in a perfect grid. Kevin tells me this is the village of Marshall.

Up ahead, five young boys stand in the middle of the road. One of them wields a tree branch like a switch at the ankles of his friends, who holler as they jump the whip when it comes their way. Aside from these boys, the village looks abandoned.

The kids dash off between the huts as we roll to a stop. The two women in the front seat climb out and laugh with each other. They are happy the ride is over, happy to be home, but no one in the *wagi* is laughing. Marshall seems a cold and temporary place, an orderly little town cut straight out of the bush. It was built by the government when the dam was closed; its residents are from one of the villages that was swamped when the waters rose in the Brokopondo valley.

In five minutes we are hammering down the Red Road again. I can't sleep. My head keeps slamming into the roof and my ankles are twisted between the men in front of me. A profound, bone-level ache spreads through my body. My joints have ceased to bend and my neck refuses to swivel; I feel as if I've aged thirty years in the last few hours.

And then the passengers suddenly come alive. The *wagi* fills with chatter as the Maroons take up old debates and arguments they'd abandoned hours ago. They speak in Saramaccan, the Saramaka language, a singsong creole more musical and less aggressive than Sranantongo, and although I only understand bits and pieces I can tell the people are excited. The *wagi* still tosses back and forth, the red wake of the dump trucks still obscures the road, the pylons still stalk the jungle. But none of this matters now. A trip that usually takes two hours has taken four and a half, but our journey is nearing its end.

A few minutes later we take another road that winds into the bush. "Home sweet home," says Kevin. We pass an elderly man riding a bicycle and the driver honks in greeting. Two young men on an ancient dirt bike knife out of the bush in front of us, their unbuttoned shirts flapping behind them like sails. We follow these two up a small rise and then leave the road for a smaller path. My first ride on the lunatic Red Road is over. Soon we are surrounded by the huts and cookhouses of Balingsoela.

3
—

HANSÉ-PAI IS DEAD

Kevin leads me to his home past huts with zinc roofs and wooden siding. The ground is dry, red, and cracked, the top layer of bauxite eroded away by decades of rain. Women and children stare out at us from darkened doorways. A lone chicken leads the way.

In five minutes we reach a cluster of huts. One of them, a small A-frame with zinc roofing on top of thatch, has a modest porch. This is where Kevin drops his bags. As he unlocks the door, I see a young girl in a yellow tank top slipping into the hut across from us.

The house is tiny and dark. In the front half there is just enough room for a small desk, a few shelves, two plastic chairs, the Peace Corps water filtration system (two white buckets, one on top of the other, with a filter between), and a small fridge. The walls are sparsely decorated with a map of Suriname and a calendar. A camouflage hammock hangs from the rafters, making the room seem even more cramped.

In the back is Kevin's bedroom: a double bed cocooned by a green bug net, a few more shelves. His bedroom door, a sheet of yellow fabric with red ovals reminiscent of Mick Jagger's mouth, is thumbtacked to the wall.

"It's small," says Kevin, holding back the sheet. "But at least I'll always be able to say I slept in Ashley Judd's bed."

Seven years ago, *Marie Claire* magazine paid for the actress Ashley Judd to travel to Suriname to become a Peace Corps volunteer.

The actress had apparently toyed with the idea of joining the Corps in her early twenties, but instead chose to move to Los Angeles to pursue an acting career. The editors of *Marie Claire* thought it would be fun to send Judd, who had recently been named one of the fifty most beautiful people on earth, to rough it in the wilds in service to both her country and her country's ravenous appetite for wild publicity stunts.

The Peace Corps operates in 139 countries. Its mission is to provide trained workers free of charge to developing nations in need of skilled labourers, with the underlying goal of promoting peace and understanding between America and the rest of the world. Volunteers serve for a minimum of two years. They are stationed with governments, NGOs, non-profits, and everyday citizens and work in areas as diverse as education, health care, information technology, and agriculture. New PCVs have little say over where they are stationed. At best, they can specify which continent they would prefer; then it's up to management to place them according to their skill sets and language capabilities. None of the volunteers I have met in this country requested to be sent to Suriname. In truth, none of them had heard of the place before.

Apparently, Ashley Judd actually chose to come to Suriname. This house was built for her, as was the double bed. She stayed for between one and six weeks, depending on the source, foregoing the usual three months of cultural training and spending much of her time escaping mosquitoes "the size of her hand." Then she returned to Hollywood and starred with Tommy Lee Jones in *Double Jeopardy*. But her influence on the insular world of Balingsoela endures.

"When I first got here," says Kevin, "every time my cell phone rang my neighbours just assumed it was Ashley. I'd be talking to the director of Peace Corps about my leaky roof, and they'd all think I'm talking to the star of *Kiss the Girls*." Kevin laughs as he tells me this but also shakes his head. He knows the villagers are mostly

teasing him, that they aren't as naive as he's just made them sound. What vexes him is that he joined the Peace Corps to get away from American pop culture and now here he is, deep in the jungle, literally shacking up with it. "A woman on the far side of the village still has a picture of herself with Ashley, taken in front of my house. She comes over and shows it to me from time to time. She thinks I'm Ashley's husband. She thinks I'm in all her movies. She also thinks that someday Ashley will return."

As Kevin checks the refrigerator for mould and sweeps two week's worth of rat shit from his floor, I inspect the colourful map of Suriname on his wall. It's a simplistic drawing that might have been painted by a child. The coastal floodplain is brown, the southern savannahs are yellow, and everything between is green. Improbably blue rivers divide the top half of the country, then slim in the south and split into tributaries, making a filigree of blue across the green.

I hear footsteps shuffling toward the house, then a woman's voice pierces the darkness of the hut.

"*Hansé-pai! Hansé-pai!*"

"Right on schedule," says Kevin, opening the door. Outside, a woman wrapped in a blue *pangi*, the traditional cotton skirt, stands just beyond the porch. At her side is the little girl in the yellow top.

"Everyone here calls me *Hansé-pai*," Kevin tells me as we walk outside. "It means either handsome or stupid-looking. I'm not sure which."

"*Hansé-pai*," yells the little girl happily as Kevin emerges. When she sees me, though, she shrinks behind the folds of her mother's skirts. Before Kevin can say anything, the woman launches into questions that tumble out of her mouth in thick, rapid explosions, her voice coming from somewhere near the back of her throat. Kevin translates.

"You have come back?" she yells.

"Yes," says Kevin.

"You are not dead?"

"No."

"I thought *Hansé-pai* is dead."

"No. Not dead."

"Sooo."

"How are you?" asks Kevin.

"Have you brought us any meat?"

"No. No meat."

"Sooo."

The woman's name is Rita and the girl in yellow is her daughter, Safira. They are Kevin's closest neighbours.

Now another woman approaches, a skinny woman with her hair wrapped in a huge bundle on her head. This is Nena. Nena tells Kevin she is happy he is not dead. Then she points at me.

"This is my friend," says Kevin.

"Why is he here?"

"*He go koiri.*"

Kevin's favourite word in Sranantongo is *dalek* and mine is *koiri*. It means to wander, to explore, to visit, or simply to walk. It is an aimless word, like the Australian walkabout. There is no goal to a *koiri*. This one word contains the essence of travel.

"How old is he?" asks Nena, looking me up and down.

"I'm twenty-eight."

Both women burst out laughing.

"Why are you laughing?" I ask.

Both women, in fits, reach up and grab their hair. Rita screams something completely unintelligible over and over. Then Kevin translates.

"It's your hair," he says. "They can't believe you have so much grey hair."

Rita screams.

"Rita is almost forty and doesn't have one grey hair."

Rita screams again.

"She wants to know how many children you have."

"None."

Both women stop laughing. A serious look comes over their faces. Kevin had warned me of this conversation.

"Are you married?" asks Nena.

"No."

"You don't have a woman?"

"Yes, I have a woman. But we are not married yet."

"Why not?"

"Because she is in Canada and I am here."

"Then you don't have a woman."

"Yes, I do."

"Would you like a woman?"

"I have a woman."

"Would you like another?" Nena asks, her gold front tooth flashing.

I twist my hair between my fingers. "I am too old for you."

Rita laughs again and steps backward, onto Safira's foot. Safira yelps and runs into her house. Nena just stares at me.

I try to avoid her gaze. In the fading sunlight, I notice my arms and legs are coated in bauxite.

"We should wash," I say.

"Yes," says Kevin. "*Wi go wasi.*"

"*Oh, yu go wasi,*" says Rita.

"*Oh, yu go wasi,*" says Nena.

In seconds, Rita has retreated to her house and Nena is gone.

We wrap ourselves in towels and walk down to the creek. We pass well-kept thatched-roof wooden huts and smoking cookhouses, the women boiling rice, the men sitting on their stoops chatting and sharpening their machetes. Everyone we pass, without fail, asks us where we are going. Their tone is neither friendly nor demanding, as if the questions are simply part of a script the whole village keeps to when two men wearing nothing but towels wander past their houses. Kevin repeats the same words over and over – *wi go wasi, wi go wasi, wi go wasi.*

We hop through a small swamp, following a narrow path into the bush. Kevin tells me he saw an anaconda here last week, stretched out lazily across the trail. The villagers never come down here without a machete, and rarely come at sunset. Most prefer to bathe in the Suriname River on the other side of the village.

The creek is not much more than a metre deep and I'm surprised by how cold the water is. On the far bank, tangled bushes hang over the water, the gaps between the leaves turning yellow then orange with the lowering sun. In the jungle's depths the insects begin to chorus. We wash quickly, scraping the bauxite from our skin. I watch the red foam form around us then drift away, the lifeblood of the Surinamese economy riding the lifeblood of the Surinamese people.

Back at Kevin's hut, we find five young children playing on his porch. Four of them are battling for bum space on two plastic chairs and the other is climbing one of the roof posts. Kevin curses himself for leaving the chairs outside.

"*Hansé-pai! Hansé-pai!*" they yell when they see us coming. Kevin yells something back and the climbing child slips back to the ground. Then Rita emerges from her cookhouse, a crude lean-to with blackened zinc walls and no roof. She yells at the children and they fall quiet. The climber, a precocious three-year-old, walks toward me with his arms straight out as if he wants a hug. I grab him under the armpits and he screams in delight. Then I raise him to the roof where he'd been eyeing something. He gently pokes the carcass of a brilliant blue *morpho* butterfly.

The kids watch us cook our dinner of curried chick peas and onions. As the rice boils they teach me their favourite game.

"High five!" they yell, and we slap hands.

"*Boxstu!*" they yell, and we punch fists.

"Gimme two!" they yell, and we touch fingers.

"Fuck you!" they yell, and they give me the finger.

We share our food with Rita. As we're finishing, a well-built man approaches the hut with a big smile on his face. He wears jeans, an orange NDP T-shirt, and a green ball-cap. He speaks perfect English.

"*Hansé-pai*," he says happily. "You're not dead."

"Nope," says Kevin. "This is my friend Andrew. He works for Conservation International."

The man shakes my hand. "Do you know Mark?" he asks.

"Mark?"

"You work for CI?"

"I did. For a bit."

"Then you must know Mark. And Fritz. You know Fritz?"

"Fritz? No. Who's Mark? Who's Fritz?"

"Fritz worked for Mark."

"Mark and Fritz? No."

Kevin stares at me, mystified. I stare at the man, shaking my head. He stares back and shakes his head, too, his smile shrinking. Then I realize who he's talking about.

"Oh, you mean Mark Plotkin!"

The man stabs his index finger at me. "Yes!"

"And Fritz von Troon!"

"Yes!"

Kevin leans back in his chair.

"I've never met them," I say. "But I'd like to. Is Fritz still alive?"

The man leans over the porch railing. "He lives on an island in the middle of the sea," he says quietly. "But I have his phone number. I'll get it for you."

In the chronicles of Surinamese conservation, two Americans stand out. One is Russell Mittermeier, the world-renowned tropical biologist and charismatic president of Conservation International. The other is a man named Mark Plotkin.

Mittermeier and Plotkin go way back. Both played crucial roles during the formative years of CI in the late 1980s. Both have been hailed by *Time* magazine as Heroes of the Planet. And it was Mittermeier who first introduced Plotkin to the wonders of tropical fieldwork, when the legendary Harvard biologist invited the ambitious young Harvard Zoology Museum staffer to accompany him to the jungles of French Guiana in search of black caiman. For Plotkin, the trip was the realization of a long-held dream, first inspired by a night course he had taken at Harvard taught by the father of modern ethnobotany, Dr. Richard E. Schultes. These two experiences convinced Plotkin, then a college dropout, to dedicate his life to the study of rainforest peoples, the healing properties of rainforest plants, and the medicine men, or shamans, who concoct and administer these medicines.

Of all the books on Suriname I read after leaving my monkeys, Mark Plotkin's *Tales of a Shaman's Apprentice* did more than any other to inspire my return. The book describes more than a decade of scientific and spiritual adventure in the jungles of Amazonia as Plotkin lived and studied with various shamans. From the Trio medicine men in the south of Suriname, he learned the recipe for *curare*, the plant-based poison Amerindians smear on the tips of their arrows, a recipe long considered the Holy Grail among ethnobotanists. While investigating the origins of a village-wide curse, he learned the cure for black skin worm, swollen testicles, and "electric eel disease" from a Wayanan witch doctor in Guyana. And with the shamans of the remote Yanomamo tribe on the Brazil–Venezuela border, Plotkin participated in hallucinogenic rituals, taking blast after blast from the shaman's pipe, feeling a "great peace" wash over him as the boiled sap of the wild nutmeg tree – the "semen of the sun" – took him to an alternative reality. Throughout his journeys, Plotkin took part in numerous hallucinogenic healing ceremonies, all of which challenged his rationalistic view of existence.

When Plotkin arrived in Suriname for his first solo research trip in December 1979, he spent two weeks studying the botanical knowledge of the Maroons. It was with the Maroons that Plotkin first witnessed the remarkable healing power of rainforest plants when he sat through a frenzied healing ceremony for a sick young boy. Plotkin's guide on that trip was a man widely considered the best jungle guide in the country, a young Maroon who knew everything there was to know about the forest – a man with the inimitable name of Fritz von Troon.

"I was expecting a blond Dutchman with a pageboy haircut, short pants, and wooden shoes," writes Plotkin. "The door opened and a large black man entered. He was six feet tall and moved with a grace that belied his extremely muscular build. His hair was short and he wore a mustache and huge muttonchop sideburns. As we shook hands, the guide asked, 'How do you do?'

"A Maroon who spoke English! I was ecstatic.

"'Fine, thank you, and how are you?'

"'How do you do?' he repeated.

"I asked, 'Do you speak English?'

"'How do you do?' he replied."

Language difficulties aside, Plotkin and von Troon became fast friends and worked together whenever Plotkin returned to Suriname, and with the publication of *Tales*, von Troon became something of a national celebrity. When I first read about him, I was taken by the image of this man traipsing through the forest with Plotkin and teaching him everything his people had learned about the plants of their new home. The idea that he is still alive and that he lives nearby – that I might meet this character from Plotkin's book – is completely thrilling.

Every night, an hour after the sun goes down, the power comes on in Balingsoela. As part of the dam agreement, Suralco provides a small amount of electricity to the villages along the Afobaka Highway, many of which were relocated to the road when the dam was closed. Balingsoela, the southernmost village in Brokopondo to avoid the flooding, has always been here. The dam is just a few kilometres to the south.

For the Maroons, electricity brought minor improvements: house lights for the few who could afford bulbs, music for the one or two in each village who could afford a stereo, refrigeration for those who

could afford a fridge (the Peace Corps and *winkel* [store] owners).
But over the years, as the villages have drawn more and more on
this "free" power (that is, power given to them in return for having
their homes and ancient villages drowned), the supply has become
unreliable and routinely gives out.

"Unreliable electricity is no electricity," says Kevin. "The power
gives out five or six times a night here. It's ridiculous."

As we're finishing dinner, the light bulb in Kevin's front room
snaps on and his fridge starts to hum. Light trickles through the
wood-slat walls of the neighbouring houses and we hear the pounding rhythm of Surinamese reggae from somewhere in the village.
It's beer o'clock. Kevin locks his door, the children on the porch
scatter, and we stumble through the dark to the village *winkel*.

Outside the store, a gang of men have gathered around a pul-
sating speaker beneath a palm-frond shelter, smoking joints and
drinking Parbo. As we approach, a few of them wave coolly to
Kevin, welcoming him back from the city and the brink of death,
without sacrificing an iota of machismo. None of them pay any
attention to me.

The *winkel* is owned and operated by a young Chinese couple,
who stand behind thick security bars, prisoners of their cache of
rifle cartridges, batteries, sunglasses, diapers, menthols, and bite-
size caramels. Neither speaks a word of Sranantongo or Saramaccan
and their faces are set in a permanent fear-gaze – eyes wide, mouths
slightly agape, hair standing on end. The couple are likely here to
stay for a few years, make a few bucks, and then retire to the outskirts
of Paramaribo, but my first impression is that they don't have any
plan at all. They seem completely free of contingencies, as if they're
watching a prison riot and have decided it's safer to stay in their cell.

We buy two *jugos*. Back at Kevin's hut, the children are still there
and some of their mothers have arrived to collect them. We drink,
watch the kids ignore their mothers and chase each other across the

path. I retire to my hammock, drunk and numb, a long day of travel taking its toll.

I wake to the sound of sweeping. In the pre-dawn darkness, someone is shuffling slowly around our house, scraping the ground with a broom of twigs. I peer out. It's a neighbour, clearing the debris that has fallen during Kevin's absence, making sure no snakes are lurking beneath the leaves.

We cross the village to the Suriname River. Every woman we pass greets us in traditional Saramaka style.

"*I wéki nô?*" they say. You have woken.

"*Mi wéki-o!*" we respond. Yes, I have woken.

"*Um fa yu wéki?*" they ask. How did you wake?

"*Mi wéki wansé wansé,*" we say. "*Um fa yu wéki?*" I woke well. How did you wake?

"*Mi wéki wansé wansé.*" I woke well.

"*So.*" Good.

"*So.*" Good.

The men greet us, too, but often at the end they add, "*Mi wéki taanga.*" I got laid last night.

The morning greeting is of great importance. Villagers follow the script to the letter and take offence if you pass by without engaging them in it. If it is still early, and few people are awake, the ritual feels almost like worship – you and I have woken in this wild place, you and I have survived the night. But once everyone is up and the sun has risen, it begins to feel like an obligation. As we walk, Kevin and I are continually engaged in at least three simultaneous greetings. Kevin handles them effortlessly, speaking in rounds and maintaining eye contact with everyone at once. But I manage to insult the oldest woman in the village – a woman with cat whisker tattoos on her cheeks – by asking her if she had sex last night.

It takes us half an hour to reach the other side of Balingsoela. Past the village shrine and beyond a tall archway of dead palm leaves a steep hill leads down through the trees to the river. Along both sides of the path, leaves have been strung in a makeshift fence.

"The pathway of the dead," says Kevin, directing me to the left-hand side. "When a villager dies, their body is carried down this hill to the river. The men walk on the left and carry the body above their heads, and the women walk on the right. They do it for villagers and they do it for large snakes. When someone kills a boa or an anaconda, the corpse is carried down this hill and taken across the river to be buried."

At the bottom of the hill, the Suriname River slides northward. A white foam on the surface is evidence of the water's recent passage through the gates of the dam. On the shore, bare-breasted women lean over buckets, scrubbing dishes and gossiping. A few metres into the water, a small group of children playfight as they brush their teeth, slipping beneath the surface and suddenly reappearing, spewing arcs of water. As we approach, the younger women pull their *pangis* up over their breasts and the children stop and stare. It's Monday, and these kids should be in school by now, but the teachers haven't arrived from Paramaribo and so school is cancelled.

We wade in. The water is murky, thick with plants and shreds of garbage. As I play with the kids, Kevin explains my presence to the women. I hear the word *koiri* repeated over and over before I dive in. One of the kids, a young boy of six, joins me underwater to shake my hand.

Beyond the children the current is strong. The kids do not follow me out this far, and when I drop beneath the surface for a moment I reemerge ten metres downstream. The women order their children to the shore and Kevin tells me not to go farther.

"People die here all the time," he says. "The villagers say there's *ogri* in the water."

"*Ogri?*"

"Yeah. Like Loch Ness or something."

We wash. A plastic bag floats past on its way downstream. I pluck it out of the water and peer inside at the remains of a boiled chicken. Then I hear a loud puff of air. Its source is a small, dark object in the middle of the rushing river. Every few seconds it lets off a massive exhalation, like a sea mammal coming up to breathe. It swims straight for us, perpendicular to the current and leaving a tight wake.

I struggle back toward shore. Halfway there I look back. The creature is gone.

"Where'd it go?" I ask.

"It went under," says Kevin.

I feel the riverbed under my toes, stand up and brace myself against the current. Then the beast emerges with a huge blast of air just behind me, and I can't decide if I'm relieved or disappointed.

The man's skin looks dry though it's wet. His dreads are wet though they look dry. His teeth are yellow and cracked-out, and around his neck is a strap of leather carved with the word "JAH."

"*Tabak?*" says the Rasta.

"No," I say.

"No *tabak?*" he says.

"No. No *tabak.*"

The man stands up straight and the river streams off his wasted body. He is just skin and bones and livid scars. He stands there for a moment, his chest heaving, whispering something to himself over and over. Then he drops back into the water.

"Fucking junkies," says Kevin.

We wait for him to surface, but he doesn't.

I carry the bag of chicken bones up the pathway of the dead. When I reach the archway I realize I've broken some kind of spiritual pact, reversing the Saramaká order of life and death by bringing the bones of a long-dead chicken back inside the village. I consider dropping the bag in the bush at the side of the trail, absolving myself of the crime, but decide that might be worse. I'm about to mention this to Kevin when an old man appears in front of us.

I hide the bag behind my back as we run through the greeting.

"*Um fa yu wéki?*"

"*Mi wéki wansé wansé. Um fa yu wéki?*"

"*Mi wéki taanga.*"

"*Soooo.*"

"*Soooo.*"

The man smiles up at me from beneath the brim of a mesh ballcap. His face is a maze of wrinkles. Kevin introduces me to Mangay, the *basia* or assistant headman of Balingsoela.

Mangay reaches out and grabs my hand. As he shakes it, he grips his own elbow with his other hand, a sign of respect. I do the same.

"What do you think of our village?" asks Mangay.

"Beautiful," I say.

Kevin and Mangay speak for a while. The Basia's eyes dart constantly from us to the huts surrounding us, but he listens intently and laughs when Kevin says everyone thought he was dead. The two seem to be very close. Mangay wishes me a pleasant visit. We continue walking, the bag of bones behind my back. Then Kevin tells me that death was waiting for him when he first arrived in Balingsoela.

Maroon cultures throughout the Americas are still, at their core, African cultures. Their rituals, customs, music, and folk tales are still rooted in the motherland, as if their ancestors had never been violently ripped from their homes in the Gold Coast or Kongo and shipped across the sea. The Saramaka are proud animists in heart and spirit, dancing to their own *apinti* drummers and praying to their old rainforest *obias*. They are both profoundly at home in the jungles of Suriname – the land their rebel ancestors fought for and won more than three hundred years ago – and profoundly out of place with respect to history and geography.

In Saramaka cosmology, the past is a potent force. Maroon ancestors possess immense power – a deed done a century ago is just as capable of affecting the present as is the action of a living, breathing soul. Richard Price, who with his wife, Sally, is the leading expert on Surinamese Maroons, writes, "Not only does each misfortune, illness, or death stem from a specific past misdeed, but every offence, whether against people or gods, bears someday its bitter fruit." In Saramaka, this karmic belief translates into an passionate veneration of the dead.

Six weeks ago, on his first morning in Balingsoela, Kevin woke to a group of women howling outside his hut. Having just arrived in the village, Kevin was frightened by the commotion, and it took him a few minutes to work up the courage to open his front door. When he did, he spotted Rita among the hysterical women. Between her cries, Rita told Kevin that the village was in deep mourning. The night before, Basia Mangay's mother had died.

Maroon societies in Suriname are matrilineal – Maroons inherit their kinship ties, social stature, land claims, and clan membership from their mothers. Almost everything a woman owns – from the houses they live in to the canoes they paddle to the protein in their diet (excluding their staple rice) – is provided by men, many of whom have more than one wife. But it is the women who "make the lineage." Traditionally, each Maroon village consists of a loosely related family, people descended from one or a few women. The death of an elderly woman – especially one with many children, and especially the mother of a headman – is a potent, traumatic event.

The keening continued all morning. In the afternoon, Kevin joined his fellow villagers at the ancestral shrine. Over the next few hours, the women cooked a feast of rice, fish, and boiled cassava on an open fire while the men built a casket for the deceased. Kevin helped bash a few nails and drill a few holes; he was struck by how carefully the men measured and sawed the wood. He also felt profoundly privileged. He had only just arrived in Balingsoela and here he was helping build a coffin for the most revered woman in the village.

The next day, Kevin went with the men across the river and deep into the jungle to the Balingsoela cemetery. After much debate, the elders agreed upon a suitable spot and the men began clearing the underbrush with machetes. The burial plot was measured and sketched into the earth. Basia Mangay arrived with eight live chickens bound at the feet, strings of fresh river fish, two cases of Parbo, bottles of rum, and cartons of cigarettes. Then the work began. For the next two days, more than twenty men laboured at the grave site, chipping away at the rocky earth with pickaxes and shovels, pounding on *apinti* drums, singing traditional Saramaka songs of life and death, and drinking copious volumes of booze. Meanwhile, the younger men each fashioned a phallus from tree branches and argued over whose wooden penis was the finest. Throughout the excavation, a brass bell was kept ringing, each man taking his turn with the mallet and providing a rhythm for the diggers. Kevin was overcome when someone handed the bell to him.

"I had just spent a decade living in a VW bus, touring as a roadie for Phish," he says. "Then I came here and found the coolest, most sweetest place in the world. The people loved to make music, to party, to celebrate life. It was a funeral, but it felt like a welcoming, too."

Once the hole was finished, the boys threw their wooden members into the grave, the elders poured rum over them, and the crew returned to the village. The women greeted them at the shore with an enormous feast. Then the villagers walked back up to the shrine, where the body of Mangay's mother lay inside her coffin.

Kevin is not sure of the significance of what happened next.

"A little ceremony took place," he says. "Two men lifted the open casket onto their heads and walked it up to each villager. When the casket came within a foot of their face, each villager extended a hand and gently pushed the pallbearer away. The drums were still beating and everyone was drunk and singing. The sun had set and a huge bonfire had been lit. Then it was my turn. I was nervous and drunk. I could smell the woman's decomposing body. So I just did what everyone else did. I pushed it away."

The next day, after an all night party, the casket was nailed shut and walked down the pathway of the dead to be taken across the river. There was no room for Kevin in the boat, so he stayed onshore with the women to watch Basia Mangay's mother begin her journey to the land of the ancestors. As the boat cast off, the women began howling hysterically and fainting, just as they had three days before.

The Saramaka belief in the power of ancestors means existence for them is a constant struggle to right the wrongs of the past. In a world where all evil has its roots in human deeds, the pursuit of justice becomes an obsession with vengeance and retribution. An unexpected death may be interpreted as murder from beyond the grave. Through divination, the guilty are revealed, with the descendants of a murderer held accountable for the crime. Then revenge is sought.

But there is an uplifting corollary to this. If every action will someday bear fruit, bitter or sweet, all Maroons have the potential to wield enormous influence on their community. Alive or dead, every Saramaka has the power to shape the world.

Kevin rolls his backpack tight and shoves it inside mine.

"We'll have to sneak out," he says. "I've already been gone for two weeks."

"Maybe I should go alone."

"No. I'm coming." He crams a towel in and fastens the straps. "I've lived here for six weeks and I still haven't seen the lake."

I peek out the window. "We're clear."

Kevin locks his door. "Boats leave at noon. We'll spend the night in Baku, with Dara, and be back before anyone notices. If anyone stops us, let me do the talking."

We head for the clearing where the *wagi* dropped us off yesterday. In five minutes we reach the village road, free as birds.

"Unbelievable," says Kevin, as we make our way up a short hill. "Our journey has been blessed. We shall give thanks with cold libations at the universe's earliest convenience."

Then we hear the children.

"Shit," says Kevin.

Over the hill they come, the lucky ones riding rickety Dutch bicycles, the unlucky sprinting down the road like jackals, outsized limbs and shoulder blades jerking in all directions. The ones at the back are the first to see us. Their yells are quickly taken up by the pack.

"*Hansé-pai! Hansé-pai! Hansé-pai!*"

There are ten of them. They speed up as they reach us, pedalling and pumping as hard as they can, showing off their newfound brawn. The boys are young and small but sure of themselves, full of pride and early hope, the next generation of tough Saramaka, a legion of ancestors behind them. I laugh but Kevin doesn't. The boys have caught us red-footed and will surely spill the beans to their mothers.

4

—

THE RIVERBONES

We walk south, following a long bend in the Red Road. Far away, almost at the horizon, is what looks like an unnaturally straight line of trees. It takes me a moment to realize what it is. The dam. We walk faster, the sun scalding us. When the road finally veers east, the dam is less than a mile away, its concrete walls barricading the lake, the old river, inside. We cross a steel bridge, beneath which the newly freed Suriname rushes north to the Caribbean. Among the reeds, a man sprays insecticide from a bottle strapped to his back.

Sweat pouring, the sun at its height, we flag a passing car to take us the rest of the way. We climb a steep hill, the small car struggling against the grade. Through a break in the trees I catch the glimpse I've been anticipating for five years. Water shimmering to the horizon. A dugout canoe creeping north like a black sliver. And across the entire lake – the one Saramaka still call the river – what looks like a forest of grey toothpicks is spread wide upon the surface.

We have reached Afobaka. A way station for Brazilian outlaws, a boat launch to the Sara Kreek goldbush. A liquored-up lily pad from which to leap into the lake.

Our taxi drops us at the shore. In the water, scores of dugouts bob in the shallows. All of them have names painted on their hulls – *Lobi Libi*, *Little Flexi*, *Almighty* – and most are laden with gasoline barrels. An emaciated twelve-year-old boy hops from gunwale to gunwale, wrestling the trigger head of a gas hose. Anxious boatmen pace the shoreline, money purses strung around their necks, rainbows of fuel at their feet. An army of *wagis* sits with hatchbacks open, their drivers scrubbing the bauxite from their frames. A collection of ancient trawlers, rusted tugboats, sun-bleached barges, and abandoned dugouts lies heavy in the uphill reeds, a good few metres above the waterline.

Thick hoses snake their way uphill to the gasoline registers, which are housed in a padlocked, rusty cage. Three Maroons sit inside, counting money and drinking Parbo and yelling instructions to the balancing boy. Men ring the money cage, all of them smoking Morellos or sipping rum from plastic cups. They cling to the grill like confused inmates, handing wads of cash through the links. A little way off, the next generation of Surinamese gas jockeys sit beneath ragtag shelters and glare at me, their fists pumping to Saramaka reggae. Uphill in the doorways of derelict buildings and beneath a lilting water tower sits the previous generation, men who remember when the flooding began.

We find a driver headed to Baku. He smiles when we ask him the price and drops it by half when I respond in Sranantongo. He says he will leave in fifteen minutes. Kevin walks uphill to the *winkel* and orders some *bami kip*. I watch our bags.

A small Hindustani man approaches. He wears a yellow polo shirt that hangs down to his knees, which are knobby and scarred and make me feel suddenly hopeless. His name is Suki.

I soon learn that Suki works in the bush around the lake, operating pumps for sand and stone miners. His family lives in the city but he can't find work there, so he comes out here for a few months at a time. He is not happy.

"Diwali day not good," he says. "I lonely man, Diwali day. Boss hate Hindus. Boss no like me. Boss want fire me but no one know my job. Boss has *bigi* family, *bigi* money. Boss no work for money. Boss go sailing Saturdays."

At this, Suki leaps up and pretends to sail, his arms out to the sides, his eyes closed to an imaginary breeze. He sways from side to side and for a moment looks supremely peaceful. His enormous shirt is transparent where it clings to his sweaty skin.

"Boss an ugly man. Ugly lady. Ugly children. Boss go fishing Sundays."

Now Suki opens his eyes wide and casts an imaginary fishing rod into the lake. He yanks on his imaginary line. "Boss say Suki drink *furu sopi*." Too much booze. Suki yanks again and pretends to reel in a big one. "*Ma* Suki *no abi* money for *sopi*."

Suki sits down and picks up the invisible fish wriggling on the end of his line. He bends over at the waist, hiding the fish in his lap, and gives me a quick smile. "Suki bushman, *sabi*?" With a violent twist, he yanks the imaginary hook from the fish's mouth and cracks the animal's phantom neck. "No one know bush like Suki."

He goes quiet and stares out over the water. His right hand still grips his catch. His left hand traces the scars on his knees.

Suki sits with me until Kevin returns. Then he races off down the shoreline, his lunch rolled up in the extra fabric of his shirt. I look around and realize the boatmen are all staring at me. A few of them are laughing, and when I catch their eyes they either look away or laugh harder and give me a thumbs-up. Afobaka is not as transient a place as I'd thought. The boatmen are the resident labourers, the cagemen are the resident sharks, and Suki is the resident fool.

An hour later we board the boat. Storm clouds have amassed in the east and a midday gloom has descended. I sit in the bow, behind a Maroon teenager wearing the shorts and jersey of the Surinamese national soccer team. We push off and the engine grunts to life. As we pick up speed, water begins trickling in beneath the patchwork of sheet metal hammered into the hull. In Suriname, this is how you know your journey has begun – the water is seeping in, your possessions are getting wet, and the boatman behind you has begun bailing. If you are lucky you are in the bow, where the only evidence that you are halfway to sinking is the constant sound of a plastic plate scraping water from the stern.

As we emerge from the bay I look back toward Afobaka. The port looks tiny and transient again, but it is not our distance that makes it so. It is the great dam, now visible to the west of the village, that dwarfs the shoreline and everything on it. Staring at this dam, I am entranced by its golden proportions and mathematical walls, which seem somehow older than the jungle and water that push against it. It looks as if it has been carved from ancient stone, from bedrock that would surely reveal itself if the forest and the lake suddenly vanished. It is an illusion, of course – the dam is only forty years old, while the jungle and water beneath it are ageless. But it is a healthy, man-made illusion, one which confirms all of our basic misconceptions in one easy scene.

The illusion does not last. As we leave the bay and enter the lake proper, we pass among the trees, where illusions immediately fade.

From here to the horizon, the crowns of fully grown trees rise lifeless out of the water. Hollowed tree trunks and leafless branches reach stubbornly skyward, bleached by the wind and rain. Some are more than ten metres high; others barely break the surface. Some still bear branches; others are just stumps above the water. All are a uniform, ethereal grey. This is the canopy of the old forest, the jungle that was flooded when the river rose forty years ago. We are floating through the dead canopy of a drowned rainforest. These trees are the bones of the river.

Our boatman follows the river's old route to the far end of the lake, the treeless path that winds south through the water. Far below,

the remains of forty-three Maroon villages lie scattered on the lake bottom, huts and Moravian churches and *gado-osos*, sites of animist worship, haunting the jungle floor. Among these ruins, an antique railroad is said to run the width of the lake, a remnant of an ancient gold rush. A man in Paramaribo says he's seen it. He'd been diving in search of gold and found ghost tracks in the murk.

Without warning, it starts to rain. The boy in front hands me his cassette tapes, which I put in the waterproof pocket of my rain jacket. We hunker down as the storm descends. We can barely see but still our boatman keeps us on track. He knows the route by heart, having driven it daily for years. He stands resolutely in the stern, blindly conducting us through the riverbones, water pouring off his ball-cap and pooling at his feet.

The storm ends quickly with a burst of sunlight and the lake flashes to life. In all directions the sun reflects off the legion of dead mahogany and dead fig and dead ceiba. The effect is blinding. We veer east down another tree-lined corridor – Sara Kreek, an old tributary of the old river – and aim for a thin line of smoke on the southern shore. The boy in front emerges from beneath his towel and taps a frantic rhythm on the bow, singing reggae into the wind.

An hour later, we turn east again and enter a large bay. The boatman slows as the dead trees crowd in and our path becomes less clear. On the southern shore, the zinc roofs of a few huts glint in the forest. Kevin tells me this is Lebidoti. Baku is five minutes from here.

Now the bones are so close I can reach out and touch them. My fingers slip across their smooth skins, each one uniquely polished, wizened by the tropical air. I rap them with my knuckles and hear their dull notes. I grab them and pull back, expecting them to snap, but they hold firm, strong as the day they died.

5
—

THE VENETIAN CARNIVAL

We pull into a tiny bay, where a group of children are playing in the shallows. It is strange to see so much life, after drifting through so much decay. A girl leaps from the bow of her canoe and waves to us from mid-air. Another girl crouches by the shore, carefully descaling a fish. Two boys walk on water and race their boats on strings.

Dara is not here. So says the girl at the shoreline, the one now gutting the fish. Her arms and kneecaps are flecked with silver.

"*Yepi-mai no dya*," she says, spilling the fish's insides. "*Yepi-mai gwe na gron.*"

"*Oten a kon baka?*" I ask.

"*Neti.*" The women will return at dusk.

Up a small hill we find Dara's hut. The door is padlocked and so is the cookhouse. We passed no one on our way here.

"I would die for this kind of quiet," says Kevin.

We drop our bags and walk to the *winkel*. Beneath the counter, two sleeping babies are sprawled on a multicoloured sheet. Their mothers sit next to them, quietly sewing.

They ask us why we're here. We say we're here to *koiri*.

"The boats are finished for the day," says Kevin. "We will stay with *Yepi-mai* and take the first boat back in the morning." Dara, or *Yepi-mai*, is another Peace Corps volunteer, one of the few I haven't met yet.

We ask why the village is so quiet. The women laugh. The men are off hunting, they say. They have to go far to find meat. They've been gone for three days and only the women, the young and old are left behind.

One of the babies starts to cry. The baby's mother rolls her eyes and says something about men always being hungry. She hefts her child up to her breast while her friend shows me the table cloth she is sewing. On a yellow background, thin blue lines spread out from the centre and spiral in the corners. The woman is very proud and tells me she made her *pangi*, too.

News of our arrival has spread quickly and six children suddenly materialize under a nearby shelter. They hide behind each other and point at us; they break into sudden dance moves. This is the Baku Welcoming Committee. One of the boys drags a toy behind him, a small truck made from an old oil can with bottle caps for wheels. A plastic beer cup has been flattened into a makeshift propeller. As he pulls the string, two batteries rattle out of the flatbed and slowly roll away.

"*Wani-pai! Wani-pai!*" they yell at Kevin, mistaking him for Nathan, another Peace Corps volunteer who lives nearby in Lebidoti.

"No!" yells Kevin. "*Mi Hansé-pai!*"

The children escort us around the village. In the shade of huts and beneath lush mango trees, old women sit cracking peanuts into giant wooden bowls. Rounds of cassava bread line the zinc rooftops, turning to gold as they bake in the sun. Occasionally, a teenaged boy steps from between the huts and glares at us, puffing his chest and pretending at manhood in the absence of his elders.

Kevin greets one of the women as we pass.

"*I wéki nô?*" he says. The woman frowns and launches into an angry tirade.

Though Baku is in Saramaka territory, this is a Ndyuka village. Its people are descendants of the eastern Maroons who settled the

95

Tapanahoni River toward French Guiana. The Ndyuka speak a different language and have distinct customs and rituals. Speaking Saramaccan here is the equivalent of speaking English to a diehard separatist in Rivière-du-Loup. They may understand what you're saying but usually refuse to respond.

Fortunately, Ndyuka is similar to Sranantongo, so I take over and explain to the woman where we've come from. She frowns at my terrible accent and waves us away. Meanwhile, the children throw stones into the trees. A ripe mango soon thumps to the ground and the children race to collect it. "*Wani-pai! Wani-pai!*" they yell, presenting Kevin with their kill. "No," he says. "*Mi Hansé-pai.*"

One of the children walks with a severe limp and is always a few steps behind the others. Every time a mango plummets, he is the last to reach the cluster of kids, hopping around the periphery while the others dive to the dirt. When I secretly offer him a slice of the sweet flesh I see the reason for his caution. His right leg below the knee has been skinned. The wound is a long red gash flecked with black soil and pieces of grass. As he sucks on the mango meat, rivulets of pus stream down his shin.

The boy doesn't seem to be in pain. When he is finished with the mango he throws the skin at the wall of a nearby hut and hobbles after the others who are warring again. I ask a young girl what happened to the boy.

Yesterday was his fourth birthday, she says. His mother was cooking a feast to celebrate, but she'd run out of onions. She'd told one of her daughters to watch the pots while she went to fetch more. Somehow, while she was gone, the cooking oil had caught fire. When she returned, she found her hut about to be engulfed in flames. There was no sign of her daughter. In a panic, she grabbed the blazing pot and threw its contents out the window, saving her home. The birthday boy happened to be running past. His leg was smothered with a pot-load of searing hot noodles.

"What happened to the boy's sister?" I ask.

"She was whipped," she says. "No one's seen her since."

We find a beautiful young woman sitting on a plank of orange wood. She is reading a book.

"*Wani-pai*," she says to Kevin.

"No," Kevin responds. "*Mi Hansé-pai.*"

"Spoon!" says the woman. Kevin sits down next to her.

This is Esther, Dara's counterpart in the village. Esther is the president of Stichting Krin Libi, a women's organization in Baku, and is learning to speak English.

"Spoon," she says again. She points at Kevin's feet. "Sandals." She points at my hand. "Pen." Her accent is very good. "Mattress. Pot. Rice. Vegetable. Machete. Cloth. Fire." She lists off the entire inventory of her house. When she is finished she gives a huge smile, her gold front tooth sparkling. We chat for a while in broken English as the sun drops down.

"Are you hungry?" she eventually asks.

"Yes," we say.

"I am sorry," she says. "I can't cook for you."

We hear a boat approaching from the west.

"I am sorry," says Esther again. "You are hungry but I can't cook."

The women have already unloaded the boats and are washing in the lake. They are indistinguishable from each other, caked head to toe with mud and ash, dark silhouettes in the early evening light. They have spent the day slashing and burning a new garden plot. Behind them, the riverbones have doubled in size, their shadows stretched long on the surface of the bay.

"*Hansé-pai!*" yells one of the blackened figures.

"*Yepi-mai!*" yells Kevin.

"Is that fresh meat?"

"Fresh Canadian meat!"

The figure says something in Ndyuka to the other women, who laugh and splash her. A rapid-fire conversation ensues, punctuated by the uniquely Ndyuka expression, "*Eeeyah!*" The women speak all at once, battling to be heard, their voices ringing in the dark.

Dara's face is masked in black muck as she steps onshore. Only the white legs poking from beneath her colourful *pangi* tell that she is not a young Ndyuka woman returning from a long day of work in the bush. "I love this place," Dara tells me as she unlocks her front door. "Even when the men are here, Baku is 90 per cent women. I've only been here two months but I've already got twenty new sisters and three new mothers. I feel like I was born here."

Dara lights a lantern and we bring our bags inside. A woman walks past and says something softly. Dara runs to the door and replies with perfect diction.

"*Eeeyah,*" says the woman.

"*Eeeyah*," says Dara.

Dara hangs her wet clothes, washes her face, and unlocks her cookhouse. Within minutes she is cooking rice on her stove and talking with two more women who have joined her. Laughter and singsong voices drift into the hut as Kevin and I hang our hammocks.

"How did Dara learn Ndyuka so fast?" I ask, as I tie a slip-knot for the fifth time.

"No idea," he says. "She was fluent by the end of training."

We join Dara in the cookhouse and she tells us the good news. "Esther can't cook but her sister can. She's such a sweet woman. She brought you some fish soup and cassava bread."

"Why can't Esther cook?"

"She's in the moon-house."

A woman's role in Maroon society is both defined and limited by her ability to reproduce. Her uterine fluids are believed to be exceedingly dangerous and capable of contaminating the ritual power of men. "Female pollution" makes women ineligible for most positions of political and ceremonial power. It also means a menstruating women must observe a number of prohibitions and largely sequester herself. Observation of these exclusions is mandatory.

"Menstruating women must walk around, rather than through, the palm-leaf structures that mark village entrances," writes Sally Price in her book *Co-Wives and Calabashes*, "and in each village there are particular areas and paths that are banned to them as well. [They] may not sit on stools, touch small babies, burn a garden site, plant crops, wash clothes at designated stones in the river, hand anything to a man, skin game or cook food for men, travel in a canoe with a man, or carry water that will be used by others."

Every village has at least one menstrual hut – in Ndyuka it's called the *mun-oso*, or moon-house – where women sleep and cook for the duration of their periods. The huts are typically located on the periphery of the village to underscore the isolation of the women.

I ask Dara if the taboo includes Americans and she laughs.

"Technically, yes. But the Peace Corps told us we could choose."

"So what did you choose?"

Dara doles out the soup and hands me a bowl. "I do it. It makes me crazy, and a bit angry, but I do it." She cracks the cap off a Parbo and takes a long swig.

She says that most female volunteers lie and say they've had a Depo-Provera shot that lasts an entire year. But not Dara. She is the first volunteer to be posted in Baku and her goal is to integrate herself into Ndyuka culture as deeply as possible, to build trust between the villagers and the Peace Corps so that future volunteers will find traction for their development projects.

"These are my sisters," she says. "I have to do it. Every month, I live in the moon-house for three days."

A violent thunderstorm crashes around us as we're finishing dinner. I spend the rest of the evening drinking and listening to Kevin and Dara gossip about their Peace Corps friends.

There is something of a carnival to the Peace Corps experience. It involves donning costumes (*pangis* and *kamisas*), wearing masks (mud and ash), subsuming one's cultural identity. Volunteers – like party-goers in Venice or Rio, or at Burning Man in the Black Rock Desert – disguise themselves to varying degrees, aware of the extent to which they have become cultural, spiritual, and sexual chameleons. The richness of their experience, the wisdom goes, depends on their ability and willingness to transform. And the same, of course, can be said of travellers.

But there is something disconcerting in the degree to which Dara has immersed herself in the culture here and become, by her own admission, Ndyuka. Her perfect accent, her expert adoption of the sense of humour, even her style of walking smacks of insincerity even as she sincerely tries to fit in here. She is not Ndyuka, she will never be Ndyuka – she knows this, of course. She is pretending to be Ndyuka.

Her act is at the heart of the Peace Corps experience. Compared with Kevin, for whom trust was built in Balingsoela by previous volunteers, Dara has shed all traces of her previous self in pursuit of her new role. And in order to continue living this way Dara must constantly deny the undeniable: that she is not Ndyuka. She is an American, and in two years she will be free to go home while her new "mothers and sisters" remain in their one-room huts on the shores of a haunted lake.

As I lie in my hammock, the *jugos* empty and the lightning flickering through the thatch, I feel I am making a cameo in a strange, post-colonial theatre, where thespian emissaries from a more lavish stage spread the impossible gospel of patriotism and peace while swearing off the trappings of the society that taught them these ideals.

Such profound contradictions go hand in hand with development work the world over; in no way do they cancel out the good works. But negotiating these psychological entanglements once the party is over must be difficult.

This is also true for the traveller who longs for home.

Because the man who longs for home is not a traveller. He is a worse actor than the development worker, a worse liar than the politician. He moves through his days like a water spider, skimming the surface of everything with no curiosity about what lies beneath. The traveller whose mind is elsewhere learns nothing. The traveller whose heart is elsewhere feels nothing. These people are neither here nor there and should just go home. Their *koiri* are wasted on them.

I worry I am becoming one of these people. I dream about Emma every night.

The boat picks up Kevin and me in the predawn cold. Two teen-aged boys sit in the bow as the boatman steers us slowly through the night, winding his way through the riverbones that clutter the bay. He picks them out at the last second with the weak beam of his flashlight and wrenches the boat around them. As we pass, dead limbs lean into the boat and scrape its hull. We doze off to their hollow voices.

Soon we are motoring north. The wind is cold, the rising sun a slow fire in the east. We shiver beneath our towels as the boys in front check their dreads in a pocket mirror and sing Saramaka reggae songs. They are on their way to the city for the first time in months, on the prowl for trouble and women. They might return when the money runs out. They might never see their villages again.

In less than an hour, the Afobaka Dam looms on the horizon and all the old illusions return. In the early light it seems pink and harmless, above suspicion, an innocuous concrete divide. The river-bones are the colour of innocence, too, blushed by the sun, traces of lifeblood climbing their haggard limbs. It's hard to tell who is performing and who is wounded this morning – until, that is, the sun climbs higher. Then the bones lose their colour and I peer below the surface. Forty years ago, as the waters rose, the bloated corpses of dead Maroons bobbed to the surface of this lake. Exhumed from their jungle graves, they wore the horrific masks and tattered costumes of the deceased, flushed from the land of the ancestors by the trapped waters of their home river.

This lake has always been a bad actor, unable to conceal what lies beneath. Promising wealth and progress, delivering death and decay. In 1999, Suralco decommissioned its aluminum smelter at Paranam, the sole reason this reservoir was created in the first place, because the dam had become too clogged with silt to power it.

6

—

LA SEULE MINE D'OR

I've heard of a group of French Canadians living deep in the jungles of Brokopondo. Today, Kevin and I are going to see them.

After another hellish ride from Balingsoela on the Afobaka Highway, our driver turns onto a slim dirt road flanked by thick jungle. We pass a small building surrounded by ten-foot fencing – "Explosives," says our driver – and eventually the jungle thins out. Through the sparse trees I glimpse towering piles of grey rubble and enormous dump trucks with tires the size of our *wagi*. Beyond this roadside fringe, the jungle has been razed.

The driver drops us off in front of an unremarkable single-storey building coated with a fine layer of white dust. "Camp David," he says. "Hurry, you are late." Inside, we are ushered into a frigid conference room with a group of Surinamese bankers who have just arrived from Paramaribo. On each seat is a yellow hard hat and a pair of safety goggles. At the front of the room is a team of mining engineers in jeans, hard hats, and flannel shirts.

We have been invited by one of Kevin's friends to tag along on a public relations tour of the Gross Rosebel Gold Mine, owned and operated by the Canadian company Cambior. Based in Montreal, Cambior is one of North America's top ten gold producers. As a Canadian, I feel duty bound to visit its Suriname mine.

"Welcome to the only gold mine in Suriname!" yells Martin Bergeron, a friendly Québécois who wears many hats at Rosebel but is technically head of catering. "I hope you enjoyed the ride!"

The bankers laugh at this. I laugh, too, but mostly because of the incongruity of a French accent deep in the jungle. Bergeron frowns.

"Cambior is the only company in Suriname that invests in the Afobaka Highway," he says. "Over the last ten years, we have invested more than $200,000 in its maintenance."

The only person who isn't laughing is a woman across from me, who sits with her forehead pressed to the table, waiting for her car-sickness to pass.

Bergeron waits for us to settle down and then apologizes that the customary corporate executive isn't available to greet us. Then he invites us to help ourselves to the spread of cream pastries at the front of the room. At this, the sick woman across from me leaps from her chair and scurries out of the room. We hear her muted gags as she fiddles with a doorknob in the hallway. We laugh. Bergeron smiles toothily. The tour is about to begin.

The legend of El Dorado – of a golden empire deep in the jungles of South America – is enjoying a renaissance. In the forests of Guyana, next door to Suriname, the world's largest gold rush is raging. Driven by gold's soaring price, thousands of small-scale miners, or porknockers, are pouring into the Guyanese interior, desperate for a way out of the crippling poverty so widespread in the old British colony. They have been followed there by multinational gold companies, Cambior among them, just as intent on striking it rich.

The Guiana Shield, the ancient sheet of Precambrian rock that lies under Suriname and Guyana, has proven to be rich in gold. Suriname's saving grace so far has been its small population and its relative affluence compared to Guyana. But the fever is rising. Historically, gold was mined by local Maroons in Suriname as a sort of emergency fund, a supplement to hunting, fishing, and agriculture. Nowadays, in the face of rampant inflation, thousands of Maroons are turning to the artisanal gold mines of Suriname's Greenstone Belt for their main source of income.

The eastward progression of the legend of El Dorado is in keeping with its ancient origins. The legend began in 1535, with Spanish reports of a Colombian king, "the gilded one," who would coat himself in oil and roll himself in gold dust. As the Spanish observed more and more Amerindians in possession of gold objects, the story evolved into a tale of a bottomless gold mine, which soon became the driving force of Europe's conquest of the New World.

Over time, the mythical treasure trove moved gradually eastward to whichever region was least explored. The Spanish mounted hundreds of expeditions from the coast of Colombia, Peru, and Ecuador, most of which disappeared without a trace. By 1600, the Spanish had given up. But soon after, Sir Walter Raleigh took up the challenge in the name of Queen Elizabeth, making two unsuccessful voyages into the Orinoco Basin in Venezuela. Raleigh was convinced the city of gold lay deep inside the "Large, Rich and Bewtiful Empyre of Guiana," the jungles of the Guiana Shield, on the shores of the legendary Lake Parima. When he returned home empty handed for the second time, poor Raleigh had his head chopped off. But the legend lived on.

With our woozy comrade back in her seat, Martin Bergeron launches into a brief history of Gross Rosebel.

Gold was first discovered here in the 1870s. Since then, the promising Rosebel concession has passed back and forth between private companies and the Surinamese government, with international giants such as Placer Development, Grasshopper Aluminum, and the Canadian junior Golden Star Resources each holding a stake at one time or another. "Rosebel has always been enormous potential," says Bergeron, his accent thickening with excitement. "Now we are realize this potential."

In 1994, Golden Star and the Surinamese government signed the Gross Rosebel Mineral Agreement, the first of its kind in the nation's history, granting Golden Star exclusive exploration rights to more than seventeen thousand hectares. Soon after, Cambior, which had been peripherally involved as a potential partner, acquired a 50 per cent interest. Exploration went ahead, and its results suggested Rosebel might be the most promising gold concession in the entire Guiana Shield. The price of gold, though, did not cooperate, and the project was stalled until construction of the mine became economically feasible. They waited a long time – nearly ten years – until prices rebounded and began the meteoric rise to today's record levels of more than US$900 an ounce. In 2002, Cambior bought out Golden Star and became 95 per cent owners of Rosebel Gold Mines N.V., with the government of Suriname owning the other 5 per cent. In 2003, construction began, and on February 11, 2004, the "only gold mine in Suriname" officially opened.

What Bergeron doesn't mention, of course, is that the 1994 Mineral Agreement was enormously controversial. Many Surinamese believe the government was crazy to hand over 95 per cent of the country's most valuable goldfield to a foreign company. Supporters, mostly those with ties to the Venetiaan government, argued the revenue generated by the mine would help lift Suriname from the tragedies of its recent past and usher in a new era of foreign investment – a noble, legitimate goal. But to those with a good grasp of South American history, the Rosebel deal merely signalled a continuation of the status quo, of rich, powerful companies taking advantage of poor, powerless nations.

The Rosebel controversy goes well beyond the sphere of economic philosophy. In the southern block of the concession lies the Ndyuka Maroon village of Nieuw Koffiekamp, one of twenty "transmigration" villages built when the Afobaka Dam was closed. Leaving their ancestral lands was an enormously traumatic experience for the more than six thousand Maroons who were forced off by the rising waters. The move has left a legacy of social and economic struggle throughout Brokopondo and Lower Saramaka, but for the eight hundred residents of Nieuw Koffiekamp, the Gross Rosebel Mineral Agreement of 1994 added salt to these wounds.

Surinamese Maroon communities have long considered themselves to be "states within a state." This interpretation stretches all the way back to 1760, when Suriname's first rebel slaves reached a peace agreement with the Dutch government, finally winning their freedom and bringing an end to nearly a century of jungle war. Since then, the legal and cultural autonomy of the Maroons has been recognized, if passively, first by the Dutch and then by Surinamese governments. Over the last fifty years, this autonomy has slowly eroded as the forests, bauxite, and gold in Suriname's interior have begun attracting the attention of foreign capitalists.

Under the International Convention on the Elimination of All Forms of Racial Discrimination (which Suriname signed in 1984), the Maroons qualify as Tribal Peoples and should, therefore, enjoy absolute ownership and control over the resources found within their traditional territories. The Convention also states that lands that have been unfairly usurped in the past (for example, the entire Brokopondo reservoir) should be repatriated. In law and in practice, though, the Surinamese government ignores its international commitments. The nation's constitution does not recognize past treaties between the Maroons and the Dutch. The Peace Accord of 1992 says that the state "shall endeavor that legal mechanisms be created, by which citizens who live and reside in a tribal setting will be able to secure a real title to their respective living areas." The nebulous and often contradictory language of the Accord has allowed the Surinamese government to renege on the promises.

Under domestic law, all land in the interior of Suriname is classified as *domainland,* or privately owned state land, an oxymoron if ever there was one. This means more than fifty thousand Surinamese Maroons are considered nothing more than "permissive occupiers" of state land, with no ownership or resource rights. This allows the government to sell off huge swaths of Suriname's interior to forestry and mining companies without consulting its inhabitants.

This is why the Maroons of Saramaka were not consulted in the early 1960s when the dam was built to power the Paranam aluminum smelter – they were simply told to get out of the way or drown. And this is why the villagers of Nieuw Koffiekamp protested in 1996 when, two years after Venetiaan signed the Mineral Agreement, officials from Golden Star announced that Nieuw Koffiekamp would have to be relocated, for the second time in thirty years, to make way for an industrial mine. Almost overnight, a number of small-scale gold miners from the surrounding communities invaded the concession, claiming their moral and historical right to be there. But when Golden Star threatened to pull out of the deal because of the protest, the Surinamese government capitulated, stationing a permanent paramilitary force on the mine site and threatening to attack the "squatters" if they refused to leave.

There were stories of security personnel strafing the miners' camps with live ammunition. In response, the Maroons set up roadblocks on the Rosebel access road. Gaama Songo Aboikoni, the paramount chief of Saramaka, travelled downriver to intervene. But even he failed to resolve the conflict.

In 1996, Ronald Venetiaan lost the presidency to Jules Wijdenbosch. Now the conflict took on a decidedly more obdurate air, with the government acting as if the relocation of the village were a foregone conclusion. The Maroons "have to decide whether they want development or whether they want to remain backward people living in the bush," said the minister of natural resources. Desi Bouterse – the former military dictator, convicted drug smuggler, human rights violator, and sworn enemy of Suriname's Maroons during the civil war – served as a state advisor for Wijdenbosch on the Rosebel project. When a representative for Nieuw Koffiekamp returned from a meeting in Washington, D.C., Bouterse publicly threatened to kill him.

The conflict at Rosebel has still not been resolved. The mine has been built, the gold is leaving the country, and Nieuw Koffiekamp still sits in the centre of the concession's southern block.

As we nibble on cake and fiddle with our hard hats, Bergeron turns the floor over to three mining engineers. One of them sketches a crude drawing on the chalkboard of gold-veins reaching like bones into the earth. Then he pummels us with an avalanche of impressive statistics. Only one in one thousand gold discoveries actually becomes a mine; Cambior invested US$159 million in Rosebel between 1994 and 2004; 66,000 tonnes of rock are mined here every day; 20,000 tonnes of this is gold-containing ore; daily gold production at Rosebel is 30 kilograms; 30 kilograms of gold is worth approximately US$473,000.

An hour later, the speeches are finished and we shuffle outside to the stultifying heat, where we pile into white buses and drive through a moonscape of dust and broken rock. We stop at a gargantuan Caterpillar 777 dump truck so the bankers can take photos of themselves dwarfed by the tires, each of which costs $12,000. Then we reach one of Rosebel's six open-pit mines, and my sense of scale vanishes.

I look over the lip of the mine and am immediately disoriented.
The depth and size of this hole in the ground, encircled as it is by
remnant forest, fills me with a deep apprehension – something like
fear but closer to panic. I have to squat to quell my vertigo. I marvel
at the shorn rock, the colourful splash of naked minerals, the ter-
raced walls that resemble giant staircases. As our tour guide hollers
statistics through his megaphone, we gawk at the trucks toiling at the
mine bottom, reduced to the size of ants by the pit, a kilometre wide
and five hundred metres deep.

We return to the buses and are next driven to the grinding mill,
where a conveyor belt delivers boulders of raw ore from the mines.
Here, the rock is ground by tumbling steel balls in a cylindrical mill
until the ore is fine enough for the gold to leach out when immersed
in a slurry of water, lime, and cyanide.

The actual gold content of the ore is very low – typically less than
0.001 per cent – and the most economical way to extract it is to

expose it to a complexant, a chemical that allows gold to dissolve in solution. Cyanide is a very good gold complexant and has been the leaching chemical of choice in the industry since the technique was refined in the 1970s. It is also toxic even in low concentrations.

The gold-cyanide complex is pumped to leaching vats for the final stages of extraction. Crushed coconut shells from the Philippines are added to the slurry. The coconut binds to the gold, allowing it to be precipitated out. But we have one more stop to make before visiting the refinery. We drive past the mines again, gaping at the vast pits on our left and the gigantic piles of waste rock amassed on our right.

"It's true!" yells our guide through his megaphone. "We are making holes in the ground. But we are also making mountains!"

What he doesn't say is that Cambior is also responsible for the second largest cyanide spill in gold mining history.

In August 1995, a faultily designed tailings dam at the company's Omai Mine in Guyana collapsed, spewing an estimated 3.4 billion litres of cyanide-laden waste water into the Essequibo River, Guyana's main source of fresh water. At the time, Omai was the second largest gold mine in South America and Cambior's flagship asset, an immense deposit that seemed to prove the legend of El Dorado once and for all, turning Sir Walter Raleigh into a doomed prophet. Here was his kingdom of gold, deep in the "Bewtiful Empyre of Guiana." And here was the accompanying curse.

Leon Carrington, an eyewitness to the spill, told a reporter, "The cyanide waste came like a great brown slick covering the water. . . . There were dead fish everywhere. It made me sick to see it." Residents downriver from the mine soon complained of miscarriages, skin rashes, and eye infections. Cambior's offices in Georgetown were swarmed with protestors. A number of damage lawsuits are still pending.

Gold companies are well aware of the dangers, both real and political, of their reliance on cyanide leaching. But there are no international regulations governing their use of the poison. In 2000, after a huge cyanide spill at a mine in Romania, the United Nations Environment Program drafted the International Cyanide Management Code, a series of regulations for the safe handling, use, and disposal of cyanide in gold operations. The Code is purely voluntary, allowing companies to choose to comply with the recommendations and become signatories; to date, Cambior has not. It is essentially operating on an environmental honour system.

Although no one died and the long-term effects of the spill are unclear, the failure at Omai was a public relations nightmare for Cambior. The company lost the trust of locals, which to that point had been tenuous at best. The public had been up in arms about the mine from the beginning, believing their government had been swindled into accepting only 5 per cent ownership – a nearly identical deal to the 1994 agreement between Golden Star and the Surinamese government.

Today, Rosebel has supplanted Omai as Cambior's most important asset. The company has spent the last decade rebuilding its reputation since the disaster in Guyana, and the tour I am on is part of that mission. But it is unclear whether anything has really changed. The company's website claims, "Cambior has leveraged its knowledge and expertise gained over 13 years of experience with its Omai operations," but there is no mention of any improvement in the design of its tailings dam. And on the same webpage is a worrying fact: the operation at Rosebel "is very similar to" the one that failed so catastrophically in 1995.

Our buses arrive at the Rosebel tailings pond and the driver lets us out at the shore, in front of the large pipe that brings the cyanide-laden waste from the refinery. Through a number of holes drilled in its side, a dark-brown sludge noisily spews into the pond.

"You would have to drink four litres of this stuff to have a 50 per cent chance of dying!" our tour guide bellows.

The group goes quiet. Kevin leans over and asks me if eight litres would do the trick. I laugh and megaphone-man frowns at me. Trying to justify a lake of cyanide in the middle of a pristine rainforest must be a mining engineer's worst nightmare.

"A rabbit is not going to drink four litres of something like this," he shouts over the roar of the pipe. "And I don't say this as a Cambior employee. I say this as a Suriname man!" He laughs heartily and stares me down. The bankers chuckle. There are no rabbits in Suriname.

A breeze picks up and I feel a cool mist on my skin. I take a few steps back from the gushing pipe.

Our guide continues, "When exposed to sunlight, oxygen, and water, cyanide breaks down completely in three days. It disappears without a trace!"

"But you're always pumping new cyanide in, right?" I say.

Silence. A few of the bankers ease away from me.

"Like I said," yells our guide. "I've seen ducks swimming on this pond!"

The rest of the tour is conducted in Dutch.

Our next stop is the refinery, where gold is extracted from the toxic slurry of ore, coconut, and cyanide. We climb a steep metal staircase, past men in chemical frog-suits, an emergency eye-wash station, and numerous skull-and-bone poison signs. We stand high above the leaching vats, gigantic barrels filled with churning liquid ore. Just as I ask which vat contains the Philippine coconut shells, there's a sickening yelp and a crash behind me.

The woman who'd been carsick has fainted. She narrowly missed falling over the railing into a vat filled with cyanide. Our guide immediately radios for the medic. The woman's friends rush to her side. The rest of us are quickly escorted back down the stairs.

Ten minutes later, a security guard arrives on foot, dressed in dark blue and shouldering an ancient rifle, followed by an old Red Cross ambulance. A medic jumps out. He wears a red T-shirt with a cartoon basketball player slam-dunking the ball and the words "GAME OVER" stencilled on the back.

"Where?" he asks.

"Up there," we all point.

The woman soon regains consciousness and is led slowly down the stairs. We applaud as she is helped into the back of the ambulance. The tour continues.

A half-hour later we reach the Control Room. Monitors showing live video feeds from the mill are stacked on the wall. On a large desk, two computer screens display colour-coded schematics of the entire mill. A man sits quietly in front of these screens, observing the blips of colour and occasionally jotting notes.

I'm impressed by the ingenuity of the mill design, and I begin to sympathize with Cambior, if only slightly. Their bumbling efforts to turn public opinion around against unimaginable odds now seems almost endearing to me. Slowly, this tour has taken on a modest, engaging sort of charm.

Then the door to the control room opens and a short, thick man enters. He wears a long ponytail and his arms are covered in tattoos; he is a caricature of a Hell's Angel. As our tour guide continues with his script, tattoo-man walks straight to the phone and hurriedly dials a number.

"Yeah, this is Johnny St. Clair," he says under his breath. "Yeah, that guy we sent over to you for those shots? Yeah, well, he just came back, and he's in quite a lot of pain and can't stop urinating."

My incipient sympathy for the company evaporates. According to international human rights treaties which Suriname has signed and the Peace Accord of 1992, the land of Gross Rosebel belongs to those eight hundred residents of Nieuw Koffiekamp. These rights should trump any company's right to mine there, no matter how cash-strapped and desperate for development the government of Suriname might be and no matter what the domestic law might say. To put it in terms Cambior or IAMGOLD, its owner since November 2006, might understand, this is the raison d'être of international human rights treaties: to provide a counterbalance to corrupt or irresponsible governments, to protect everyday citizens from multi-national companies that hide behind outdated and unjust legislation instead of performing their own due diligence. One hundred and fifty years ago, Mark Twain reportedly dubbed a gold mine "a hole in the ground owned by a liar." But even this sentiment doesn't apply here. After all, although you wouldn't know it by looking, the real owners of Gross Rosebel are the Surinamese Maroons.

We exit the mine complex through the Evacuation Assembly Area, where a young Hindustani man frisks us with a metal detector. After a complimentary buffet lunch, during which the woman who fainted returns to a round of applause, Kevin hitches a ride back to Balingsoela and I follow the bankers out to their waiting *wagis*.

As I walk past the guard booth, another thick-set man emerges with a big smile on his face.

"You are the Canadian!"

"Yes. *Salut.*"

"*Salut!*" He shake my hand excitedly. "What are you doing here?"

"*Koiri.*" My explanation for everything. "What are you doing here?"

"I live here six months." The man continues shaking my hand. He is badly sunburned.

"What do you do here?" I ask again.

"How are Les Habitants?" he asks.

It takes me a moment.

"I don't know," I say.

"Why?"

"I was down here when the hockey season started."

"Oh."

The man frowns. He stops shaking my hand but still holds it tightly. The *wagi* driver honks his horn.

"*Tabernac!*" I say, wrenching myself free.

"*Tabernac.*" The man's smile slowly returns. "*Oui. Tabernac.*" He looks lost, or lonely, or both.

I climb into the *wagi*. We pass a sign instructing us to "Have a Nice Day, Drive Safely, Protect Our Environment." Then we settle in for the long ride to Paramaribo. It's only when the banging of the Afobaka begins that I remember Fritz von Troon – his friend in Balingsoela never returned with his phone number.

7

WOODSTOCK *NA BUSI*

"The New World is dying," says Jesse. "The Old World is dead. Suriname is the New New World."

Three days ago, when I got back to Parbo from Rosebel, I'd found a long phone message from Bouterse's bodyguard, my new friend Bodi. He needed a favour. His wife, Maria, was scheduled to work this weekend but she was very sick. He wondered if I might cover for her. Maria works as a chef for various tour operators, and this weekend she was supposed to travel to Tonka Island, in the middle of Brokopondo Lake, to cook for a group of Dutch students. If I were willing to chop a few vegetables and follow the instructions of a man named Gudi, I could go on the trip for free. And perhaps earn a little of Bodi's trust in the process.

So now I sit with a tour operator named Jesse on the shore of a lake riddled with dead trees, waiting for the boat to Tonka. The bus is long gone and the students are exhausted; four hours of hammering down the Afobaka Highway have taken their toll. They lounge among piles of backpacks, coolers, speaker towers, boxes of records, and coils of electrical cord, some taking naps, others casually flirting. Jesse and I sit just offshore, in the hull of an abandoned dugout named *New York*.

"I am a son of house music," he says. "But it's more than just my life. It is my message."

Jesse is tall and thin, a balding man in his late thirties who has spent the better part of two decades immersed in the European electronic music scene. He was born in Suriname, but like so many in the early eighties, his parents took him to the Netherlands to escape the violence and bread lines of Bouterse's dictatorship. Jesse spent seventeen years in the Netherlands. That's where he discovered his calling.

"Amsterdam, Rotterdam, Utrecht. Everything was so dark in the mid-eighties. The kids were dying to get out, to get away, to see the world. But we couldn't. We were all broke. All we could see was the dirt and grime, the port cities, the fake liberalism. It all seemed so unreal. So we threw parties."

Jesse became a DJ. He began spinning house records at local parties and soon made a name for himself as electronic music gained momentum throughout Europe. He played in abandoned warehouses and flooded basements and the open air all over the Netherlands. He often headlined with such house luminaries as Sander Kleinenberg.

The rave revolution has now been over for some time. Across the Western world, four-on-the-floor house parties and their various incarnations have lost all trace of rebellion and have been convincingly eulogized as a social movement. But Jesse thinks this is just another case of blind, Eurocentric thinking.

"It's not over," he says with a wry smile. "It's just moving to different places. It's been to North America. It's been to Eastern Europe. It's been to Japan. Now it's coming here. I came back to Suriname three years ago for a three-week vacation and I've been here ever since. The youth here have so much energy, so much love. They want more than what they're offered by their corrupt government. The kids here are going to change things. This is the New New World."

Back in Paramaribo, Jesse and a small group of DJs throw house nights at various music venues across town. He says the response has been slow due to Suriname's obsession with reggae, dance hall, and Western hip-hop, but he's confident his gospel of revolution-through-beats will catch on. "The love is growing," he says.

This tour to Tonka Island is part of Jesse's mission. Every year for the past three years, he and a bunch of friends have organized a rave somewhere deep in the jungle. Each year, a select group of tourists is invited to travel into the interior and party for twenty-four hours straight. Last year, the party was somewhere in the east, near Albina and the Marowijne River. This year the party is on Tonka.

"We want Suriname to become the next Ibiza," he says. "We want people from all over the world to come and party here, to dance to good music, to feel the love while surrounded by our jungles, our wildlife, our culture. This is our philosophy. *If you play it, they will come.* Like Kevin Costner, right?"

I can hear the distant hum of an outboard motor and soon a boat appears at the mouth of the bay, weaving its way through the snags. It lands with a crash and an enormous black man with waist-length dreads steps to shore. This is Bote, Jesse's partner in house. The two embrace in a bear hug. Then Bote orders the mountains of gear into the hull.

The students and I take our seats. I sit next to Gudi, in part to assure him that I am here to help – to slice onions and not to party – and in part to convince the students of this. They have paid three hundred SRD each for this trip.

Jesse shoves the boat off and leaps into the bow.

"Suriname or the moon!" he yells.

The students are from Amsterdam and have been in Suriname for three months. They are here on a co-op term with their universities, partly to gain business experience and partly as goodwill ambassadors from the metropolis. We motor down the west shore of the reservoir, the forested hills of Brownsberg rising on our right and shrouded in mist. We pass small islands of jungle and smaller islands of rock and sand. Some of the students reach out and touch the dead trees, then pull their hands back quickly as if the branches were burning hot or painfully cold.

Forty years ago, when the dam was closed and the waters of the Suriname River began to rise, the Maroons were not alone in having their lands inundated. A crisis was also looming for the rainforest animals who would surely drown or starve to death as their dens and nests and hunting grounds sank beneath the rising waters.

In mid-1964, a young Bostonian named John Walsh arrived in Suriname. Walsh was a prosecuting officer for the Massachusetts SPCA, an expert on the humane capture of animals in distress. He had been sent to Suriname by his boss, who had recently received a worrying letter from the Suriname SPCA.

A few weeks after the Afobaka Dam was closed, the Suriname SPCA wrote to every other SPCA branch worldwide on behalf of Brokopondo's doomed animals. They had asked for money to help fund a rescue operation "for the anteaters, for the deer, the tapirs, the sloths, the tree porcupines, the howler and the spider monkeys, the smaller sakis, and many more animals of the tropical rainforest, who are threatened by drowning or starving." The letter signed off with the ominous and now famous line, "Time is short and the water rises." Walsh's boss, instead of sending money, decided to send his twenty-three year old protege to see if he could help.

The result was Operation Gwamba, the largest animal rescue in history (gwamba means "animal" in Saramaccan). For eighteen gruelling months, Walsh and a crew of Saramaka men lived and worked on the rising Brokopondo reservoir, rescuing every animal they could find and setting them loose on high ground. From camps perched on shrinking islands, armed with little more than the tools normally used to capture raccoons in the suburban alleyways of Boston, Walsh and his team saved 2,104 three-toed sloths, 1,051 nine-banded armadillos, 479 red howler monkeys, 161 pygmy anteaters, 36 tapirs, and 3 jaguars, just to name the larger animals. They were modern-day Noahs, herding frightened animals with packs of hunting dogs or cornering them in trees with Ketch-alls and cat poles. It is likely that some of the trees still visible today were the sites of many such rescues, the last safe place for a drenched opossum or famished kinkajou to hide.

The story of Operation Gwamba is an astonishing tale of jungle survival, stoic suffering, and near-catastrophe. Over the course of a year and a half, Walsh and his recruits battled infestations of ticks and chiggers, angry tarantulas, aggressive pit vipers, bloodthirsty vampire bats, swarms of vicious ants, weeks of dysentery, and the constant threat of malaria and dengue fever – all while wrestling frantic and half-starved red brocket deer, giant armadillos, and well-muscled wild boars into dugout canoes during torrential downpours. The island camps soon became modest zoos, where emaciated animals were caged and nursed back to health before being released. Prison breaks were common, with thirty-three bristly porcupines suddenly turning up around the campfire. One evening, a puma appeared in their makeshift kitchen.

Meanwhile, the forest was rotting beneath the surface. The water quickly grew fetid, stocked with infectious bacteria and schools of rabid piranhas. Walsh himself had a horrific bout of dengue, developed a nasty skin fungus, contracted bat salivary virus, and dodged disaster countless times through sheer perseverance or dumb luck. He once flipped his hydroplane when he hit a caiman while racing back to camp at night. Another time, he miscalculated while stalking an anaconda and barely extricated himself from the constrictor's crushing grasp. He even saved the life of an over-tranquilized howler monkey by performing artificial respiration. His worst moment, though, came when his boat flipped late at night and he had to swim, alone, over the submerged village of Kadju. Walsh knew piranha gathered in the hundreds over old village sites. He spent agonizing hours in the water, waiting for the first *ting* on his finger or toe that would open his skin and begin the feast.

The project had its failures. Due to the speed of the rising water, Walsh and his men would occasionally arrive at small islands and find the bloated corpse of a tapir bobbing near the shore, or maggot-ridden bodies of starved tortoises scattered among the trees. But on the whole, Operation Gwamba was a remarkable success. Walsh and his ragtag crew saved more than ten thousand animals from certain death.

To those who know his story, Walsh is a hero. And since his adventures in Suriname, he has dedicated his life to rescuing animals in need. In 2002, nearing the end of a distinguished career, Walsh led a delegation from the World Society for the Prevention of Cruelty to Animals to war-ravaged Afghanistan, where he oversaw the rehabilitation of the Kabul Zoo.

It takes us an hour to get to Tonka. Near the shore, a few thatched-roof buildings back onto the forest. Bote drives the boat onto the pebble beach and we carry the gear up to a two-storey guesthouse. Inside, the students rush through the tiny rooms, hanging off the rafters to test their strength and laying claim to the best hammock spots.

Behind our building is a small village of guesthouses that seem abandoned, their walls green with mould. In the rear I find a bath-house with two flush-toilets and a sink stained black with disuse. The jungle is reclaiming these buildings, slowly enveloping them in a thicket of vines and epiphytic plants. I am walking through a ghost town, the remnants of a rainforest tourist resort that never took off.

There is one house in good condition, a white building of vaguely Dutch design, with weatherboard siding and a modest porch. In front stands an old Maroon man with an impressive pot-belly. He wears nothing but a pair of torn black shorts. On his right shoulder is balanced a long plank of wood.

"*Fawaka?*" I say, reaching out my hand.

"Good," says the man. He shakes my hand, the plank wobbling up and down.

"Can I help you with that?"

"No."

"We have come for the party."

"Yes."

We stare at each other. Now I see his eyes are partly clouded over. His face is mottled with grey stubble.

"This is my island," says the man.

"It is a beautiful island," I say.

"My name is Fritz."

"I am Andrew."

"You are American?"

"No. Canadian."

We stare at each other again. Then it hits me. *He lives on an island in the middle of the sea.*

"Fritz?"

"Yes."

"Fritz von Troon?"

"You know me?" he asks.

"Yes." I shake his hand again. "I have known about you for years."

I am speaking with most famous bushman in all of Suriname. A human encyclopedia of New World tropical botany, a modern-day medicine man and national hero, the man who helped launch Dr. Mark Plotkin on his crusade to safeguard Amazonian indigenous knowledge.

I am speaking with the elderly owner of a foundering tourist resort in the middle of a man-made lake.

"I used to work for CI," I tell Fritz.

"Me too."

"I have read about you."

"Doctor Mark."

"Yes, Mark Plotkin's book."

"You know Mark?"

"No."

"Doctor Mark is *bigi* man now," says Fritz, suddenly talking faster. "I know Doctor Mark when he was *pikin* man. I help him. Take him into bush, show him plants. *Busi dresi, sabi?*"

"Medicines."

"Yes. *Med-ee-sheens.* I work long time. I like Doctor Mark. Doctor Mark like me. Then he fire me."

"Why?"

"*Mi no sabi.* I teach Doctor Mark everything. Then, one day, goodbye Fritz." Fritz bounces the plank of wood on his shoulder.

"Maybe you can help me," I say after a short silence.

"*San yu wani?*"

"I want to see the blue frog. *Okopipi.*"

"*San?*"

"Can you take me?"

Fritz laughs. "*Okopipi* not easy. Not many people know."

"But you know." I say.

"Yes."

"So? Can you show me?"

Fritz looks down at his feet. He thinks for a long time, his forehead creased. Then he pushes past me. "I must work," he says, the wood swinging out and forcing me to duck. "Tomorrow morning, eleven o'clock," he says over his shoulder. "We go for walk in bush."

Two more boatloads of tourists have come ashore and the DJs have set to work. I sit in the shade with Gudi as the disc jockeys convert an old thatched-roof building called Watra Dagu into an electronic dance hall. They fasten strings of lights and industrial strobes to the eaves, running wires around back to a huge generator. The makeshift DJ booth is simply two turntables and a mixer on a rusted table, overlooking an outdoor dance floor of sand. The records are stacked behind the table, and on either side sit two massive speakers that will drown this little island in sound. Inside Watra Dagu, which means "otter" in Sranantongo, four more speakers are wired into the corners and coolers of beer are stacked at the back.

A cry goes up from behind the building. The generator growls and the lights burst to life. Jesse throws on a record and flips a switch, and an old house track pounds to the back of our throats. Jessie howls with delight and hugs his comrades, DJ Fabian and DJ Mario. Gudi leans over and screams something in my ear.

"Woodstock *na busi!*" he yells.

"*Ai,*" I shout back. Woodstock in the bush.

The sun sets, the beer begins to flow, and Gudi leads me to the cookhouse. For the next hour, I chop onions and garlic while Jesse ramps up the party outside. There are four cooks, plus me and an eighty-year-old Carib woman named Granny. As we work, Granny walks between us, laying her hands on our shoulders and telling rude jokes to boost our morale.

The music stops and an amplified voice howls across the island.

"DJ Mario, Number One DJ in Suriname!"

A pounding techno track. The muted screams of a large crowd. I look at Granny, who is looking at me.

"What are you doing, Sonny?" she asks.

"Cooking," I say.

"What are you cooking?"

"I don't know."

Granny laughs and grabs my knife, pushes me out of the way.

"Get out of here," she says. "I am old. I will cut smelly things."

I look at Gudi. He shrugs.

"Get out of my kitchen!" yells Granny. "You are young! Go have sex!"

Watra Dagu is packed. A haze of marijuana smoke hovers near the ceiling as the tourists dance to DJ Mario. We pound kingcans of stout as strobe lights spasm and nearby trees flash suddenly to white. The floor of our makeshift dance hall shakes and bounces with the weight of more than one hundred partiers. Terrified geckos and pissed-off tarantulas scramble to the thatch. One of the students, a girl of nineteen, shrieks and points to the roof. Nobody hears her.

Down at the beach, someone has lit a bonfire. The riverbones flicker red, as do the faces of Dutch girls and boys circling the blaze as a cold wind picks up. Mario works his way through a deep and devastating set of tribal house. Halfway through he drops the volume and Jesse grabs the microphone.

"Anyone with drugs, please report to the DJ booth. *Snel! Snel!*"

As if on cue, a bolt of lightning, a burst of thunder, a sky full of raindrops frozen by the strobes. Bote appears in the murk, wrestling an enormous canvas tent. The storm hits and our sandy dance floor turns instantly to mud. Tent abandoned, shoes off, pills down, we party outside in the rain. Fabian takes over from Mario, the Number

One DJ in Suriname, and takes us on a rollercoaster through the old days of progressive trance – rousing anthems, synthetic atmospherics that match the dramatic weather. Girls and boys pair off and disappear down the shoreline. Someone whizzes a glow-stick in front of my face. The stout goes down like water and Jesse gives me an enormous hug.

"I told you!" he yells, as the music pulses and the thunder booms. "I told you! The New New World!"

And then Fritz von Troon appears among the crowds. No one notices the old man as he climbs the stairs of Watra Dagu, squeezes through the intoxicated mob and stands in front of the massive speakers. He stares at the woofers, tweeters, and squawkers, which breathe in and out like hysterical animals. He just stands there and faces the music for about five minutes – wondering, perhaps, at the number of *apinti* drums it would take to make a racket like this. Then he turns and walks away.

A few hours later, I walk down to the shoreline through the storm. The party is still raging but I need a break, some solitude to enjoy my high. I walk far behind Watra Dagu, where the lights can't reach and the music is muffled. I listen to the raindrops clattering on the surface of the lake, each one like a sharp and silver needle through my mind.

Just before Christmas of 1964, John Walsh's chronic diarrhea developed into something much worse. Clean drinking water had been one of the biggest challenges for Operation Gwamba, as the growing lake had become "a giant smelly swamp of tannic acid from decomposing trees, spiced with typhoid fever, amoebal dysentery, who knew what all." But Walsh's problem was not the water. His joints and muscles had begun to ache terribly, and his stomach was cramping as if punctured by hot swords. Walsh had dengue fever.

In a haze of pain, Walsh asked his most trusted boatman, Wimpy, to take him to the dam. He remembers little of the trip except for one surreal moment, as Wimpy paddled the boat through the half-swamped village of Ganse.

"A church run by Bushnegro Moravian Brethren had leaned over forward and was partially floating in water that, in the late afternoon, looked like ebon ink. . . . I stood up, bent over, and looked at the church through my legs, and marvelled at the reflection. And then, dizzy, fell back down. A few birds and insects were still in the trees, and the sounds seemed louder and clearer than ever before. Maybe it was because there was no noise to detract – no motors or voices or wind – or maybe it was just that my fevered mind for some reason had heightened perception – but the bird calls seemed to scream through the stillness, and the insects sizzled, all the sounds echoing through a vast amphitheater."

The horror of the destruction at Ganse allowed Walsh to transcend his pain, if only for a moment, in a state of mild hallucination. He'd been in Suriname for six months and had already saved thousands of animals, but he had yet to fully appreciate the human cost of the dam.

"We paddled slowly among the silent, empty huts, their doors open like grotesque, gaping mouths in one last gasp of agony. One door was ajar, and, as the boat slid past, our waves rolled to the hut, into the opening, and the door opened and shut, opened and shut, thud, thud, thud."

"This *is* the way the world will end," wrote Walsh. "This is the whimper."

Soon the sun comes up and splashes the riverbones with colour. The storms have passed, the tourists have retired to their hammocks but the DJs are just hitting their stride. Jesse spins everything from techno to heavy metal to lounge to funk, even the theme song from *Inspector Gadget*. Fabian and Mario, the Number One DJ in Suriname, stand behind him with their eyes closed, beers in hand, Rasta headbands around their foreheads, nodding to the earsplitting beats.

Upstairs, I find an empty bed, right above one of the speakers. I pass out to frantic dreams of heart attacks and wild animals.

Unfortunately, Fritz is right on time.

"*Wi go* little bit *na busi*," he yells over the thumping break-beats. He points up the beach toward a stand of thick jungle. I follow him out to the shore, exhausted and terribly hungover.

Fritz tells me about the flood. He was born in the village of Kadju, on the banks of the old Suriname River. When Fritz was about twenty, the waters began to rise. The people knew it was coming – his friends had been earning good wages for over a year at the dam site, hauling stone and clearing jungle, making way for the concrete monstrosity that was rumoured to have magical properties. They'd heard this giant wall could turn rushing water into light, refrigeration, even music. It's hard to say whether Fritz's friends recognized the irony – desperate for steady work, these young Maroons supplied much of the raw muscle required to destroy over half of Saramaka, their homeland.

All over Saramaka, people in forty-three villages packed up their meagre possessions and left the land their ancestors had fought and died for. Some of Fritz's neighbours went north, to the government-funded transmigration villages, while others went south to the Upper Suriname region, where the flood waters never reached and where the river still flows. This decision – north or south – held deep ramifications. Those who fled south were moving deeper into Saramaka territory, hoping to regain a traditional riverside life for themselves and distancing themselves farther from the coast and its

Western influences. Those who fled north moved to the equivalent of refugee camps for displaced Saramaka. Here they would live near the fetid shores of the reservoir, within a three-hour drive of the capital city, surrounded by people from different villages and kinship lines, in a melting pot of Saramaka families.

I ask Fritz where Kadju was originally located.

"*Langa fara*," he says, waving his arm over his shoulder without stopping. A long way away, somewhere in the middle of the lake.

Kadju is everywhere, he says. Kadju is nowhere.

We reach the stand of trees and I follow Fritz into the darkness. We walk for a few minutes across the soft forest floor. Then Fritz stops, kneels down, and carefully pushes a pile of dead branches to one side. Beneath the debris, twelve tiny plants reach their first leaves up through the soil. The leaves are so new they glow. Fritz brushes the soil from the fragile leaves. Their radiance makes the jungle surrounding us seem old and tired.

"This is my orchid place," says Fritz, circling his hand above his head.

And now I see them. Hidden beneath piles of brush, perched in the trunk forks of slim saplings, nestled into the dark recesses of nurse logs, a garden of young orchids, not yet flowering, speckles the forest with points of emerald light.

For a few, orchids embody ecological interconnectedness because of their remarkable co-evolution with the insect species that pollinate them. But it is their beauty that captivates the many, and it is their seeming fragility that Fritz enjoys. He carefully drags the branches back over the twelve new plants, hiding them from view. The music still pounds from Watra Dagu and we can feel the beats thumping beneath our feet.

We sit on a rotten log in the middle of his garden and Fritz tells me his plan for Tonka: an education centre, a herbarium, a restaurant, a bar. He wants to bring kids here to teach them what he knows. The schools in Suriname are teaching children the wrong things, he says, city things instead of forest things. He wants the children to understand just how precious their jungle country is. "You can have practice with no theory," he says, "but you can't have theory without practice." Fritz never went to school. He learned everything he knows by walking in the bush.

And so he builds. Whenever he has the time and the wood he puts up more buildings on Tonka. He takes in tourists to pay the bills, to fund the creation of his forest school. The guesthouses are actually meant for Surinamese schoolchildren. The ground floor of Watra Dagu, now strewn with exhausted, hungover ravers, was intended as a lecture hall.

"*Efu yu no leri, yu lasi*," says Fritz. If you do not learn what you have, you lose it. "Other countries lose everything," he says, gesturing at his orchids. "But we still have."

Fritz taps my knee. The project will cost 75,000 Euros, he says. He asks me if I have 75,000 Euros. I tell him no, I'm sorry, and then he asks if I know anyone with that kind of money. I shake my head. He gives a wan smile.

"I talk about me too much," he says after a short silence. "Now we talk about you. I can't take you to see *okopipi*."

"Why not?"

"You must ask permission from government."

"I tried that. They told me no."

"Then you no go."

I've spent the last few weeks daydreaming of the great Fritz von Troon leading me into the southern savannahs, the land of the Trio Indians. I don't think he understands what I'm asking.

"Why can't *we* go?" I ask.

"Government say no."

I stare at Fritz.

"You mean secret," he says.

"Yes. Secret."

Fritz thinks for a moment.

"When?"

"January."

"I have contracts," he says. "I go *na busi* in January." He'll be in the bush.

"For how long?"

Fritz counts on his fingers. "April."

My heart sinks. I can't afford to stay here until April. I am already living off my credit cards, racking up debt. There's only so much credit I can get.

We sit for a time without speaking. The distant music has turned to reggae, Caribbean dub beats wafting through the trees.

"You want to go in secret?" asks Fritz after a while.

I nod.

"You must ask permission from the Trio Granman. The Indian chief in Kwamalasamutu. If you don't ask him, he won't let you off the plane."

"How do I do that?" I ask.

Fritz stands up and reaches into the crook of a nearby tree, where an orchid is nestled on a bed of moss. He gently lifts one of its leaves and peers underneath.

"He is a friend of mine," he says. "I will radio him for you."

We walk out of the jungle the way we came. On the beach, the sun is at its height and the stones are dry. In the shallows, the riverbones are lined up like soldiers storming the shore.

"Did you know John Walsh?" I ask as we pick our way among the stones.

"Johnnie Walsh," says Fritz.

"You remember him?"

"America call Johnnie Walsh a Tarzan," he says, leaning down and picking up a piece of driftwood. He runs his thumb over its silver surface. "We call Johnnie Walsh a Suriname man."

Jesse is the only one dancing when we get back. Fabian is spinning the records he began with yesterday, classic European progressive house, and Jesse's eyes are closed as he glides across the hillside. One hand clutches a plastic cup of rum; the other punches the air to the beat. He dances smoothly, as if the rhythm were simply an extension of his shoulders and hips. Down at the shore, the students cool off in the water or laze about in Speedos and bikinis, their pale Dutch skin turning bright pink beneath stylish fedoras. Bote treads water out beyond the tourists, his dreads splayed on the surface like snakes. Everyone is ready to leave, it seems, except the DJs. Bote yells at Jesse to pack things up. Jesse gives Bote the finger.

The boats arrive with the rain. Only now does Fabian pull the plug. He and Mario spend the next half-hour coiling cords and boxing records and shifting speakers down to the water. Jesse continues to dance, the music still hammering between his ears.

Now a woman approaches from Fritz's house, a tall Saramaka dressed in traditional skirts. She has a beautiful, full figure, tightly cropped dreads and a mouthful of golden teeth. Her name is Suzanne. She is one of Fritz's daughters.

Suzanne hands me a small piece of paper with six numbers scrawled in red ink. "My father says you call him. After New Year he will be in the city. He will help you then." She walks away. Then Jesse dances past, almost crashing into her.

"This man in *ek-sta-see!*" she yells, playfully shoving Jesse back up the hill.

I glance up at the house and see Fritz sitting on his porch, watching me out of the corner of his eye. I wave. He holds up a hand, fingers spread wide, then quickly looks away.

We ride back through a violent thunderstorm. Lightning spikes the air and the small islands we passed yesterday are gone. The students jostle for room beneath a huge blue tarp but there is no room for me or Gudi. We stare out into the grey, shivering in the downpour, neither of us saying a word.

Through the murk I can see the riverbones. Something about them has changed. It must be a trick of light, the slanting rains, my atrocious hangover, my lack of sleep – the riverbones are moving. They sway from side to side, slowly, softly, tilting to the wind as if made of flesh and skin. The last triumphant gestures of the doomed. They wave to a single rhythm, the beats inside my aching head, and

I wonder if Jesse has noticed the dancing trees. I look to the bow but can't make him out in the storm.

The *wagi* is already at the swamp when we arrive. The driver is angry – apparently he's been waiting for three hours – and before the doors are shut he's flooring it back to the highway. We are fifteen people in a van made for eight. Our gear and equipment is packed floor to ceiling. And when the driver fails to slow down for a particularly sharp turn, Granny flies out of her seat and becomes pinned between a fallen speaker and a mountain of sodden records. Jesse rises from his seat and screams at the driver, at which point the driver rises from his seat and screams at Jesse. And soon we are hurtling down the most treacherous highway in all of Suriname, our driver half out of his skin with rage, Jesse half out of his mind with booze, the sleepy students wide awake with fear, Granny pretzelled in the corner and laughing her ass off, Fabian smoking a big fat joint, and Mario, the Number One DJ in Suriname, preaching to anyone who'll listen the classic, burned-out, club-kid philosophy: that it doesn't matter if we crash and die, people, because everything up to this point has been a gift, a blessing.

"Who are we to expect the party to last?"

The party, he says, is over.

8

—

GRAND CAFÉ RUMORS

Rosebel has sprung a leak. Yesterday, workers discovered liquid cyanide escaping through a crack in one of the vats. The mine has been shut down until further notice, and although Cambior officials were quick to announce the spill has been contained within safety walls, Cambior stock has taken a hit this morning on the Toronto Stock Exchange.

Three days from now, my friend Jason will arrive from Vancouver. To kill time until he gets here, I sit at my usual table at Grand Café Rumors in the lobby of the Krasnapolsky Hotel. Five years ago, I would come to the Kras to bask in the air conditioning, drink extravagantly garnished whiskey sours, read Hermann Hesse, and write self-absorbed letters home on the deck of the third floor pool. Now I'm reading Bruce Chatwin, writing obsessive emails home to Emma, and talking with strangers about Rosebel. Everyone has an opinion on the leak and everyone thinks he's right.

Gossip is the courier of news in Suriname. With such widespread government corruption and an economy unofficially based on cocaine trafficking and illicit casino revenues, secrecy is a way of life here. As a consequence, most Surinamese believe the word on the street is as reliable as anything reported in *De Ware Tijd*, the Time for Truth newspaper. The locals even have a nickname for it – *mofo korantie*, or the Spit Press. "If you really want to know what's going on," my friend Sean once told me, "you tune into the Spit."

Word travels so fast that another friend once received an accusatory phone call from his girlfriend *during* his first kiss with another woman. The lesson of the *mofo korantie* is that truth in Suriname is more democratic than absolute. Anyone can say anything on the streets of Paramaribo, and as long as they say it with conviction and style they stand a better than average chance of being believed.

I hear it all – the leak has been contained, the leak is just a political ploy and never happened, masses of fish are floating belly-up on the Mamanari Creek, the entire Rosebel area is submerged under a sea of cyanide. Everyone I speak with admits to fearing another Omai.

My phone rings. A woman's voice asks for Albert. I explain she's got the wrong number and hang up. Then a man named Wade Campbell sits down next to me.

"Gold ain't the problem," he says. "The problem here is oil."

Wade is forty-six years old, a pudgy Trinidadian who went to high school in the Bronx, college at Columbia, and who now works for Texaco in Guyana. He is in Paramaribo for talks with Venetiaan's resource minister. In his spare time, he goes to the casinos.

"You see those protests?" he asks, referring to the march last week on the Presidential Palace. The government had just announced plans to double the price of gas. "That's because of me."

"How's that?"

"I walk into their offices, talk my bullshit, and Venetiaan's men just nod and sign my dotted line. Barely have to open my mouth."

"You think they shouldn't have signed?"

"Of course they shouldn't have signed!"

"But it's your job."

"That's right." Wade smiles. "Look at me, look at me, if I'm talking about a multi-million dollar deal, I'm going to bullshit you until I get it done. This country is built on bullshit. You can't live honestly all the way through. You'll never make it."

"You must be good."

"Brother, I ain't bullshitting you. I am the king of bullshit. That's why I love casinos. Big shiny buildings built on deception. Look at me, where you from?"

"Canada."

"Big country, friend."

"Vancouver. Toronto."

"And the rent?"

"I'm a writer."

"I got family in Rexdale, on Weston Road. The long one. North Etobicoke. Look at me. You gotta do it like Peter Jennings. He's from Canada. You gotta use answers to get more answers."

"I'm not a reporter."

"You know who's a great writer? Stephen King. Look at me. You heard of him? I read his books and can't put them down. And Les Brown, motivational speaker. He married Gladys Knight. Great writer. What are you reading?"

"*In Patagonia.*"

"Ah, the king of bullshit himself."

My phone rings. The same woman's voice asks for Albert. I hang up.

"You understand what I'm telling you?" says Wade. "Men are dogs. And not just with women. You got a woman down here?"

"Back home."

"Man, the pussy down here is all about money. Women in heat down here! I had a beautiful girl, Italian and Brazilian, but I screwed it up. You know why? Men are dogs." Wade laughs. "That's what you should be writing about. Forget the gold. Where you staying?"

"Stadszending."

"Brother, if you ever need a place to crash, I'm in Room 121 upstairs. Stop by the pool this afternoon. I'll sign you in."

Wade stands, shakes my hand, and leaves. Then my phone rings again.

"Albert?"

"No! Not Albert!"

"*Disi* Vivian."

I'm about to hang up and then I remember. Vivian. The woman in yellow from the Red Road.

"*Disi* Andrew."

"Andu!"

"Yes."

"*Mi wani kon luku yu.*"

"When?"

"Now."

As I get up to leave, I notice two Krasnapolsky security guards staring at me. One of them says something into his walkie-talkie. The other nods as I pass.

Vivian and I sit in my room in Stadszending and try to talk. She barely speaks Sranantongo and my Saramaccan is terrible. She tells me about her six sisters and five brothers in Koini Kondre, how she really wants to be a nurse but her favourite thing to do in the city is gamble. She is twenty-three years old, has two children, and is dressed beautifully in white. Her hair is pulled back in a bun with a hairclip that says "LOVE."

"Why did you call me?" I ask.

"I never speak to white man before."

She sits closer, smiles at me. One of her front teeth is capped in gold. On her thumb is an ancient band-aid. Her breasts sag with the weight of motherhood.

"What happened there?" I point to her thumb.

"Monkey bite," she says. "Do you have a woman?"

"Yes."

"You are married?"

"No."

"*San yu wani?*" she asks. What do I want?

"*San?*"

"*San yu wani?*" she asks again.

"*Mi no ferstan.*" I don't understand.

Vivian sighs. She edges closer on the bed. Then my phone rings. It's Bodi.

"You come eat tonight," he says.

"Where?"

"I pick you up. Seven o'clock. My wife wants to thank you. She cook very nice."

I hang up. Vivian stares at me.

"Who was that?"

"Nobody."

"Your woman?"

"No."

Vivian smirks. "Your woman." She stands up. "Where are your clothes?"

"Huh?"

"*Mi wani wasi yu krosi.*"

Vivian wants to do my laundry.

"Vivian, you don't have to wash my clothes."

"*Mi wani.*" Vivian grabs an empty bag from the floor. "*Puti.*"

I shrug, pack my dirty laundry, and hand it to her.

"I bring," she says. "*Moni fu taxi?*"

I give her ten SRD.

"I bring," she says.

I find Wade sitting by the pool at the Krasnapolsky, ogling the Dutch women in bikinis. The water is bright green and smells like the jungle.

"My brother, you should be back in Canada, where it's modernized. All the ATMs are broken."

"Don't tell the tourists."

"Forget tourism. One McDonald's, one KFC, one Pizza Hut for an entire country? How can they expect to have tourism when businessmen can't get their money? My mobile is out of minutes." Wade stares past me at the swimmers and smiles. "Mmm mmm . . . I'll take the one in orange."

The door behind us opens. A Creole woman in a tight miniskirt walks in, her high heels clattering on the deck.

"Baby, baby, baby!" yells Wade.

The woman stops, gives Wade an insulted look.

"No, no, no," says Wade. "I know you. You were at the casino last night."

"Was I?" says the woman. Her voice is scratchy with a cold.

"My friend, this girl looked fine. Look at me, look at me, this girl looked so good. She was wearing . . . what were you wearing, baby?"

"You don't remember?"

"Something blue."

"It was yellow."

"Right. Yellow. With straps. I even wrote my number down to give you. I ain't bullshitting. I have it right here." Wade stands, sifts through his jean pockets. "Here, here," he says, holding an old receipt out to the woman.

She steps closer. "Do you remember what I said to you?" she asks.

"Yes!" Wade laughs. His front teeth are fake. "My brother, she told me to fuck off."

"No, I didn't. I told you I love you."

The woman sits down. Her name is Sandra. She has cheekbones the size of plums, enormous white teeth and a pair of fake eyebrows sketched high onto her forehead. Both of her wrists are thick with gold bangles and she carries a diamond-studded purse. She is fine-looking from a distance but up close appears to have been hurriedly, ungraciously assembled. As she sits, she crosses and uncrosses her legs, flashing us her panties. Wade nearly falls off his chair.

I ask Sandra what she's drinking and head for the bar. As I order, an old Dutchman with a colossal red nose stares at me. A glass of white wine perspires on the bar in front of him. He wears a tight pair of surfing shorts and no shirt.

"What's she need?" asks the bartender.

"Tea," I say.

"I know what she needs," says the Dutchman. He laughs and then begins to cough. The loose pink flesh of his torso shudders.

"You OK?" I ask.

"No," he wheezes. "I'm in milk." This man owns the biggest dairy farm in Suriname.

"One thousand head of cattle," he says. "I live here one week every month."

"A week in paradise."

"A week in hell. But if I don't come, cows flood."

"Flood?"

"I tell my boys to dig ditches, canals, yes? But they don't listen. Rains come. Cows flood."

I pay for the tea. The milkman grabs my arm with a clammy hand.

"Tell her I'm next."

Back at the table, Wade is laughing.

"I'll do it right here! Right now!"

"Not here," says Sandra. "Your room."

"My brother," says Wade. "Sandra tells me no man can pleasure her. Can you believe that?"

"Your room," she says again.

"You can't come to my room. It is dangerous. I work for Big Oil. Too many secrets. Give me a kiss."

"I only kiss men I love."

"You already said so!"

"I joke."

Sandra takes a sniff from a mentholyptus stick. Wade is silent. Behind us, the old Dutchman slaps into the deep end.

"My brother," says Wade. "This young lady wants your number." Sandra is getting higher by the second. Her eyes are almost closed. "Give it to her. I gotta shake." Wade walks to the bathroom. I ask Sandra for her mentholyptus. I take a snort and it rushes to my head.

"Chinese," she says, smiling into the table.

Sandra knew a Canadian once. He was a pilot for Air France. She lived with him for two years and kissed him all the time. Then she got pregnant. Three months later, he left, so Sandra drank a special drink, a jungle drink, to kill the baby. She bled and bled. She wrapped herself in towels but the blood kept coming. She lived in her bathroom for two days. When she finally went to the hospital, a last-minute blood transfusion saved her life.

"I'm sorry," I say.

Wade comes back. "Sorry for what?"

"I have to leave."

"So do I," says Sandra, heaving herself from her chair.

"Did you get his number?"

Sandra turns to me, phony eyebrows raised. Then she turns back to Wade.

"He doesn't like me."

"Sure he does!" says Wade.

"Sure I do!"

Sandra holds her hand out to Wade. "Give me money for taxi."

"Oh, baby, I would," he says. "But the ATMs are broken. Can't even fill my mobile."

Bodi and his wife Maria eat chicken the way everyone in Suriname does. They tear off massive chunks with their teeth, chew vigorously through the gristle, suck the meat from the bones, organize it all in their cheeks, and then spit the bones back onto their plates.

After dinner, Maria clears the table and Bodi and I go outside for a smoke. We stand next to a young banana plant, the evening bees humming around the fruit.

Bodi stares at me.

"What?"

He sniffs. "Someone told me you are spy."

"What?"

He takes a long drag, holds it in. "Don't worry," he says. "I don't believe them."

"Why not?"

"Everyone knows what everyone hears." He exhales a smooth line of smoke. "And I spy for two years. You are not dangerous enough."

"Thank you. Who did you spy on?"

"Enemies, *toch*."

"You mean friends?"

Bodi laughs. A truck passes, its flatbed filled with cattle. A scooter with a full-length mirror tied to the back. The buzz of insects.

"Why do you want to know about Bouterse?" he asks.

"Because I don't understand why he's not in jail."

"He has power. People holding him up."

"What do you think of him?"

"He is good and bad, just like you and me. He helps the poor. He is the only one who cares about this country."

"Did he kill those people?"

"In 1998?"

I look at Bodi. He looks away.

"What happened in 1998?"

"Nothing."

"Nothing?"

"*Noti, toch.*"

Silence.

"Then what about '82?"

Bodi shrugs. "I was ten years old."

"You worked for him."

"Yes. But later. Only two or three people know the truth," he says, crushing his smoke out on the wall. "Some people say he was at Zeelandia. Some people say he was at the barracks. Rumours, *toch*? I hear you have good weed in Canada."

"I lived in Vancouver."

"It's legal in Vancouver?"

"We call it Vansterdam."

Maria appears in the doorway with the kids. Bodi kisses them goodnight and lights another smoke.

"Sometimes," he says, "I come home, take a shower, come out here and smoke weed. Nobody knows, only me."

"I have money," I say. "I would pay you."

Bodi smiles.

"Ok," I laugh. "I don't have any money."

"Money is easy," he says. To top up his military salary, Bodi occasionally buys a few hundred grams of cocaine, cooks it up, and sells it on the street. He buys from a dealer he calls simply Big Man.

"I told you already," he says. "I don't want to talk things I've seen."

"Then why invite me for dinner?"

Bodi nods his head and thinks for a moment. "Home-cooked meal."

"I don't believe you."

Bodi laughs again, snorts smoke out his nose.

"I don't care."

Bodi's phone rings. He walks down the driveway. Before he answers he turns and raises his index finger to his lips.

The next night, Bodi pulls into Stadszending on his Yamaha.

"You want to know Bouterse? Get on."

I grip his shoulders as we motor up Dr. Sophie Redmondstraat and head south into Paramaribo's suburbs. Soon we reach a massive traffic jam. Soldiers stand at the intersections and the streets are lined with parked cars. Bodi pulls over.

"Follow them," he says, pointing to a crowd of people.

"Where are they going?"

"NDP headquarters. Bouterse speaks tonight."

I climb off, hand him my helmet.

"Maybe you have something for me?" Bodi holds his thumb and index finger to his mouth. "For buy weed?" I hand him five SRD. "Thank you," he says, stomping his kick-starter. "I see you here in two hours."

I join the crowd. We walk along an eight-foot fence hung with massive billboards. Soon we arrive at a small break in the fence, where four teenaged soldiers stand, machine guns slung over their bony shoulders. One of them pats me down.

"When does Bouterse speak?" I ask the boy.

"Bouterse?" he asks.

On the other side of the fence, I understand the boy's confusion. This isn't a rally; this is the annual Caribbean Business Fair.

Thousands of Surinamese wander down pathways lined with booths and collapsible storefronts, barrel-voiced hawkers howling their bargains. Young families and couples lick ice creams and munch popcorn. Children jazzed on cotton candy weave back and forth like drunks. Homeless men peddle stolen watches to uniformed police officers. The Good Luck Shak and Shoes is packed. The lineups at the *panenkoeken* house are a half-hour long. Music thumps from the Boem Box stand. The whole city is here to buy something, from cars to washing machines to lighting fixtures to herbal teas.

At the Cambior booth, I speak with a young Javanese woman named Valeenee Wasimin. She stands beside a sign that reads, "Mining With The Environment In Mind." Valeenee is a PR assistant with Cambior and a fervent NDP supporter. She laughs when I mention Bouterse.

"Headquarters is nearby, but nothing was planned for tonight."

I ask her what she sees in Bouterse, how she can support someone with such a dark history.

"They all have terrible pasts. The speaker of the National Assembly was convicted of sexual harassment. *Mofo korantie* says the minister of justice was a gunrunner in Guyana, the vice-president had a drop site for cocaine in the bush. Everyone knows what the other one did during the war, and they've all got their hands in each other's pockets. But the NDP is the only party with a plan. If you had heard Bouterse speak tonight, you would understand. He's the only real leader this country's ever had."

"Will there ever be a trial?" I ask.

"A trial would go nowhere."

I smile. She smiles back.

"It's true! The only man who knows what really happened was Fred Darby."

"And Fred's dead."

"Exactly."

"What about Bouterse's men? His bodyguards?"

"They'll never talk. Men in this country are more loyal than dogs."

Valeenee speaks passionately about her country's politics, its past and its future. Then I ask her about the cyanide leak at Rosebel and her face goes blank.

"There has been no impact on the environment or local communities," she says. "The leak was completely contained."

"So what about the reports in *De Ware Tijd* about contaminated creeks?"

Valeenee puts a well-trained hand on my shoulder. "Boy, come on," she says with a phony smile. "In Suriname, if people want you to believe something badly enough, they'll say just about anything."

Bodi picks me up. He is very high and trying to hide it. When I tell him about the fair, though, he quickly sobers up.

"I see crowds," he says. "I think it must be Bouterse. It has to be Bouterse."

"Why?" I ask.

"Because no one else make crowd like that."

It's been two months since I left Toronto and Emma's emails are becoming frantic. I've been gone longer than we've been together and she's finding it harder and harder to believe I exist. I write back as fast as I can, tell her about *okopipi*, my similar need for proof of rare and beautiful things. I've had three frustrating meetings with STINASU now, and each time they've denied me the permit.

I find Sandra in my seat at Grand Café Rumors. She is dressed casually now, in jeans and a T-shirt. She looks younger, maybe nineteen. Two textbooks lie open on the table.

"Your friend is at the pool," she says without looking up.

"He's not my friend."

"Sure he isn't."

"What are you reading?"

"School." She scribbles in a bright pink notebook. "I think he's *oplichter*."

"What's *oplichter*?"

"Con man. Thief. I don't think he has room here."

"What about you?"

"*San?*"

"What do you do?"

Sandra looks up. "Wait for man."

"Which man?"

She fumbles with her mobile. "Milkman."

"The old Dutch guy?"

"Yes."

"Why?"

Sandra smiles, types something into her phone. "Pay for school," she says.

I read Chatwin for a while. A security guard walks past our table. Then Sandra bursts out laughing.

"Give me your number," she says, gathering her books and pushing herself up from her chair. "Give me!"

I write it down.

"I go upstairs," she says. "I tell Wade I need to call you. For emergency. Then he'll have to take me to his room."

Before I can say anything, Sandra is at the elevators. Another security guard walks toward her. He says something and she nods her head. The elevator arrives.

Five minutes later, my phone rings.

"Brother! Come upstairs! I'll sign you in!"

"Uh . . . is Sandra there?"

"She just left. Great personality, that girl. You hit that last night, or what?"

"Is she coming down?"

A faint splashing sound.

"Wade?"

"I ain't bullshitting. You gotta see what I'm seeing."

"I just ordered a drink."

"Alright, brother. You know where I'm at."

A few minutes later, Sandra steps from the elevator. She walks right past me.

"*Oplichter*," she whispers.

The other elevator opens, revealing Wade in a crisp grey suit with a charcoal tie. Sandra disappears into the candy store on the other side of the lobby. A small surge of fear flickers down my spine.

"Nice threads," I say.

"I'm late for a meeting," he says without sitting down. "Listen, brother, can I ask you something?"

"Sure."

"Are you fucking with me?"

"What?"

Wade smiles. "Look at me."

"What?"

"That girl's a prostitute, you know that? A whore."

"I didn't know that."

"She just asked me for a hundred dollars."

I have no idea what's going on.

"Did you give it to her?" I ask.

Wade shoves his hands in his pockets, lets out a sigh. "Where'd you say you're from?"

"Vancouver. Toronto."

"My brother, if you're trying to fuck with me. . . ."

"I'm not."

Wade grabs my book from my hand and slams it to the table. Then he sits down and shakes his head. "Look at me. I've been to Western Union three times this morning. And still nothing."

"That's . . . too bad."

"Money's gotta go through channels."

"Texaco's a big company."

"That's right." Wade eyes me, leans back in his chair. "Listen, you think you can help me out?"

"Huh?"

"I feel bad about Sandra."

And as quickly as it arrived, the fear is gone.

"I've got nothing, Wade."

"Come on, brother. Look at me."

"I'm looking."

"And you've got nothing?"

"That's right."

Wade smirks. "I don't believe you."

"I don't care."

He slowly pushes himself up from his chair. "You *should* care."

I smile up at him. "Manage to fill your mobile, did you?"

He straightens his suit. "Listen, don't fuck with me."

"But this country is built on bullshit. You said it yourself."

Wade just stares at me, absently roots around in his pockets. Rage flickers around his temples. Then, without a word, he leaves.

When I next look up a thick security guard is standing at my table.

"Excuse me, sir? Are you a guest of the hotel?"

"Uh, no. I just like the bar."

"And the pool?"

I smile, look around for Sandra but she's long gone.

"The water's a little green, but it's ok."

"I'm only going to say this once, sir," he says with a tight smile. "The pool is for paying customers only."

The next day, Vivian returns with my laundry. Then we walk down to her favourite hangout, the Tropicana Casino on Saramaccastraat.

Inside the automatic doors we find a glittering land of hope. A brand new SUV perched on a golden pedestal. Rows of slot machines, each one occupied by an entranced Saramaka woman. Banks of card tables surrounded by Chinese teenagers. Roulette wheels manned by rich businessmen, their suits gleaming in the fluorescent lights. I buy two buckets of coins and give one to Vivian to thank her for my laundry. She takes me to her usual seat at the slots, next to an elderly woman in a bright red *pangi*.

"This my job," she says absently, as she bets the maximum and thumps the spin button. Whenever she comes to the city, Vivian's brother gives her twenty SRD to spend here. "I lucky," she says with a smile.

Vivian speaks the truth. My bucket is soon empty but Vivian is going strong. Every few minutes her machine sings and spews a torrent of coins. The woman next to us sucks her teeth and moves to a machine on the far wall, but Vivian doesn't notice. She is spellbound. I buy myself another bucket but a half-hour later I'm broke again.

I wait for Vivian to finish. All around us, Maroon women from the interior are throwing their money away, money they likely earned this morning selling vegetables or fish in the Central Market. Slot machines are a gloomy enterprise no matter where they are found, but these at the Tropicana are especially depressing. The people most addicted to their lies are Suriname's poorest people, the Maroons who live in the jungle.

Vivian's last coin wins her nothing. Back at Stadszending she is quieter than usual.

"Your tooth," I say, trying to make conversation. "Why do you have a gold tooth?"

She smiles. "A gift."

"From?"

"Mother."

Vivian's eyes dart back and forth around the room. She taps her foot impatiently. Her hands, wrapped in her lap, look twenty years older than mine.

"What does it mean?" I ask. "The tooth."

Vivian sighs. "I pass test in school."

"What test?"

"Hard test. Numbers."

She looks at me, opens her mouth. "See?" she says, pointing. I lean in. The gold is engraved with a percent sign.

Vivian sighs again, looks to the ceiling.

"My son is very sick," she says finally.

"Oh?"

"He has bad kind of malaria. He come from *busi* for hospital."

"I'm sorry."

Now Vivian looks up and frowns. "I need money for hospital. Can you give?"

I desperately want to believe her. I feel a fool for doing so.

"How much?"

"One hundred and thirty SRD. For doctors."

"I don't have that much."

Vivian gives a wan smile. She doesn't believe me, either.

I go to my room, dig out my emergency cash. One hundred SRD. I return to the lounge and give it to her. I am an idiot. I am a stupid Westerner who is in too deep.

"*Grantangi fo yu*," she says, thanking me. She does not smile. Instead, she looks angry, perhaps at her situation, perhaps at me.

"I go hospital now. *Grantangi.*" I hail a taxi and help her into it.

Then I jump in another cab. I am so tired of being lied to, by Wade, perhaps by Sandra. For once I want the truth. I tell my driver to follow Vivian's car. We tail her for fifteen minutes up Dr. Sophie Redmondstraat. She gets out in front of a massive building on Flustraat.

"What is this building?" I ask my driver.

"*Academisch Ziekenhuis,*" he says. The Academic Hospital.

In the *seketi* genre of Saramaka folksongs, if a person sings that their name has been "written down in a book," they are accusing another of spreading cruel rumours about them. Grand Café Rumors, a friend would later tell me, is the unofficial clubhouse for Paramaribo's thriving community of con men.

It turns out Wade isn't particularly skilled at his chosen profession. I call the Krasnapolsky and ask for his room.

"I'm sorry sir," says the front desk lady. "But there are no guests named Campbell at the moment."

"How about Room 121?"

"I'm sorry," she says again. "There is no such thing as Room 121."

9

THE CHILDREN OF LEBIDOTI

S aramaccastraat may be the Grand Central Station of Suriname, but it is also the country's secular temple, a place of longing and, therefore, a place of worship. It is home to Suriname's black markets. Here, everything is traded: Surinamese dollars for American dollars, flakes of gold for American dollars, captured parrots, monkeys, and frogs for any currency at all. The Maroons on Saramaccastraat are seeking to trade what they already have for something they desperately want: financial prosperity, a sense of traction, a leg up on a rapidly changing world.

For well over a hundred years, Saramaka Maroon society has gradually changed from being rooted in agriculture and hunting to a growing reliance on paid work. As early as the 1870s, the men started travelling far from home, lured by the promise of employment outside Saramaka as low-paid gold miners or construction workers. They also tried their hands at drug- and gun-running or simply joined the throngs of unemployed labourers on the streets of Paramaribo. Few of these men returned to Saramaka for any length of time; most pinned their futures firmly to the coast. The same is true today.

Immediately following the Peace Accord in 1992, Suriname's government began signing concessions over to multinational mining companies and Southeast Asian logging firms. This process continues today, with Surinamese politicians selling wide swaths of land to the highest bidder in the name of "development." Government offi- cials allow these companies near-complete autonomy, often pro- viding military protection to ward off confrontations with angry Maroons, while at the same time – or so says the *mofo korantie* – padding their own pockets. These concessions now comprise more than 40 per cent of Suriname and affect more than 60 per cent of the country's indigenous communities.

Today, the bauxite mines at Moengo and Paranam are nearing exhaustion, a worrying portent for an industry that contributes more than 70 per cent of the country's export earnings. With the bauxite running out, Suralco and BHP Billiton are aggressively seeking new deposits in the remote western watershed of the Bakhuis mountains. The Bakhuis project calls for another hydroelectric dam to be built, exactly like the one at Afobaka.

In the Western worldview, Suriname is a relatively prosperous member of the developing world, those economically poor, mostly tropical and subtropical countries with significant poverty, low per capita income, and low levels of industrialization. Since the end of the colonial era in the 1950s, these countries were collectively referred to as the Third World, but this term has recently been outed for what it is – an insult levelled by the rich nations of the world that has influenced the economic policy and underlying psycholo- gies of Western governments since it was coined in 1952 by the French demographer Alfred Sauvy.

But the substitute phrase "developing world" is not right, either. It suggests that there is an irresistible future in industrial capitalism for these countries, an engine of infinite pistons that will sweep them all up and carry them into a better tomorrow. It also suggests these countries have already taken some kind of giant leap toward the future they long for – after all, they are not *undeveloped* or *under- developed* any more; they are *developing*.

The happy term "developing" ignores the cultural and social sac-
rifices that are inevitable on the way to becoming a "developed"
country – domestic control over mineral rights comes to mind, as
does the survival of indigenous and tribal ways of life. It also veils
the usual result of said development – a financially poor tropical
country with an economic infrastructure that resembles the tem-
perate rich countries of the world, thereby enabling the latter to
take advantage of the former in terms of open markets, free trade,
and resource exploitation.

The term "developing" is also faulty because it implies the
"developed" world has already arrived at some kind of pinnacle of
progress, that we have doubtless followed the right path to get there,
and that we have achieved everything that self-respecting nations
should strive for. There is an insinuation of moral, cultural, and
spiritual achievement in the term that rivals the old phrase "Third
World" for its underlying arrogance.

So, I suggest a new phrase, a term free from judgment, discrimination, and moralism. The truer word, the more honest descriptor
for countries like this, I think, is *converting*. The people of Suriname
are *converting*, slowly and steadily, to the ways of a *converted* world.

Day and night, the secular temple of Saramaccastraat is alive with
gods and goddesses: the men who rule the black markets and the
women who rule the men.

Suzanne von Troon is one of these goddesses. I can see her
from two blocks away, standing a foot taller than most of the men,
as Jason and I push our way through the mass of bodies on
Saramaccastraat. The torrential rains have turned the street to mush
and the potholes to ponds. The oranges and grapefruit for sale at
the edge of the road are splattered with mud, and so are the women
who sell them.

Jason is one of my closest friends, an engineer by training but now an aspiring photographer. Four months ago, at a party in Vancouver, I convinced him to see Suriname for himself. He arrived last night and will travel with me for a month. He wants to see the river-bones, and I want to see the goldbush, so we're headed back to Brokopondo.

Suzanne sees us coming and flashes a golden smile.

"Where do you want to go?" she asks over the roar of a passing DAF truck.

"Lebidoti," I say.

"Wait here." Suzanne disappears into the crowd. A minute later she pops up in the middle of a thick gang of men and waves us over.

She stands next to an ancient *wagi* filled with young women and old men. The van is perched on four sorry tires, all of them spares. At Suzanne's side is a Saramaka boy of about sixteen. "This is Sheldon," she says. "He will take you to Afobaka." Suzanne grabs Sheldon's shoulder, leans down and looks the boy in the eye. "*Disi man no* tourist." Sheldon nods. Suzanne holds his gaze for a moment and then releases him. "You pay what we pay," she tells us.

Sheldon takes our bags, stashes them in the back, and secures the hatchback with rope. We squeeze into the back row of seats and Sheldon floors it. Through the back window I watch it all recede: Suzanne, the crowds, the temple of Saramaccastraat. Then we turn a corner and the engine chokes. Sheldon pulls over. No one says a word. The motor, he says, is *fukup*.

We lurch through the city, picking up cargo while the engine howls in complaint. Here is the pervading psychology of the *wagi* driver in Suriname, perhaps the pervading psychology of the entire Converting World: when everything from the cooling system to the electrical system to the doors of your vehicle are held together with nothing but makeshift, last-ditch solutions – ill-fitting pipe, wires stripped of their insulation, knots of withering twine twisted around the windshield wiper – you do not drive slowly. You go as fast as possible. You throw caution to the wind in the hope that you will reach your destination – be it geographic or socio-economic – before the entire vehicle crumbles beneath you or bursts into flames.

We pull up to a darkened warehouse. Sheldon throws the van into reverse and that old circus tune, "Entry of the Gladiators," blasts from a speaker hidden in the dash. Deep inside the building, two skinny Hindustani men emerge from the shadows and strap ten sheets of zinc to our roof. When they are finished they dance together, arm in arm, two dust-covered clowns waltzing to the tune.

Back on the road, Sheldon fishtails around a hairpin bend and the hatchback flies open. We spin around and watch all of our gear tumble into the mud.

"That's not good for business," says Jason, scrambling out over the rear seats. His pack, containing much of his photography equipment, has landed in a pothole filled with mud.

An hour later and the sky over Paranam is dark. The lights are on at the Suralco factory but no one, it seems, is there. We join the Afobaka Highway and Sheldon shifts into high gear, pounding over the tire ruts and skidding around the potholes. The chatty girls in the front row fall silent and the men in front of me bounce in their seats. DAF trucks and eighteen-wheelers plow past, spraying our windows with red mud. Soon the interior of the *wagi* is lit with an eerie, bauxite light.

Sheldon screams and pulls over. He jumps out onto the road and hops around, rubbing his ass, and the girls burst out laughing. Then Sheldon lifts the base of his seat, opens a small door in the floor and clouds of steam billow out from the engine.

We pile out. The men smoke cigarettes and drink early beer. The girls toy with cell phones they've pulled from their skirts. From above, gobs of saliva rain down every minute or so, courtesy of a woman who refuses to leave the *wagi*, her head lolling out of her window. A six-month-old baby is sleeping in her lap. From between this woman's ample breasts, a cell phone peeks out like a young marsupial.

Soon a man emerges from the bush carrying two buckets of water. He gives them to Sheldon, who pours them both onto the engine block. With the first, more steam. With the second, less. We climb back in. Sheldon sits gingerly in his seat. One of the old men says something funny and the rest of them wheeze and cough.

We pass through several rainstorms, stopping to let passengers out at villages with names like Judea, Pasensi, Bigi Batra, Ottobanda. The *wagi* is half-full now and reaching new speeds. We hurtle down a hill of bauxite moguls. And then the buzzing begins.

Sheldon slams on the brakes. The buzzing crescendos to an ear-splitting alarm and we all climb out. Sheldon gives us a vaguely congratulatory look, as if we've made a valiant effort but now, finally, the engine has spoken.

"We wait," he says sadly. "New *wagi* come. Sure."

The clouds threaten to unload again as we stand helpless in the mud. The midday insect chorus rises from the surrounding bush. Finally, another *wagi* descends the hill we've just ridden. It pulls up next to us and the driver laughs at our sorry state. He spends the next five minutes teasing Sheldon mercilessly.

On the roads of Brokopondo, the status of a man's van is his main source of pride. This has little to do with the roadworthiness of his vehicle, however. If the *wagi* runs well, the owner is proud. But if it runs poorly and can still make the trip, the owner is even more proud. And pride, of course, is greatest before the fall. When his *wagi* breaks down, the driver is inconsolable.

161

Sheldon takes the abuse without a word, embarrassed and ashamed that the highway has bested him. The new driver agrees to take us to the lake. In the back, Jason and I sit on either side of an enormously obese woman whose abundant flesh floods her lap as if a human dam has burst. The walls of the *wagi* are covered with Saramaccan graffiti, crude statements about various men's mothers and sisters scrawled between patches of rust.

An hour later we find Afobaka on its best behaviour. The port is quiet, orderly, professional. The music is off. There's no one drinking. The boys are no longer acting like men. The *wagis* are lined up and sparkling white. The gasoline cage is free of hangers-on. Incoming boats kill their engines early and drift silently to shore. Even the gas hoses have been untangled. As our driver opens the hatchback and dumps our bags in the dirt, I can even hear the lake lapping onto the shore.

I find Suki, the resident fool, standing next to the rusted-out carcass of boat named *Big Business*. Suki is digging a hole in the sand with his foot. His blue T-shirt billows down to his ankles.

"Why no music?" I ask.

He points uphill to a circle of men dressed in dark green military suits. At their centre is the ranking officer, leaning over a map spread on the hood of his jeep. He wears a gold-trimmed police cap and his jacket is weighed down with medals, crests, and ribbons.

"Disco music!" yells Suki.

The District Commissioner of Brokopondo – the Disco – is paying a visit to Afobaka.

At the water's edge sit two freshly painted dugouts with Surinamese flags fluttering in their stern. In of one of these sits the Disco's boatman. His name is Alfred. When he's not working, Alfred likes to fish in the lake. He lists off his favourite species: *tukanari, pirana, jogu, kwana, songeh, bongoni, kobee, anyumara, pataka.*

Two military trucks arrive. Their doors open and a platoon of soldiers empties onto the beach. Suki cowers behind me.

"Why is the army here?" I ask Alfred.

"We go Sara Kreek. Disco wants to talk to the miners."

The soldiers walk past us to the boats. They are dressed in gorgeous camouflage. On their backs, small rucksacks. In their breast pockets, packs of Morellos. Around their waists, water bottles and handguns. Over their shoulders, semi-automatic rifles and machine guns.

"Is this a raid?" I ask.

"No," says Alfred. "We go to talk."

"To talk?"

Alfred rubs his thumb and forefinger together.

"Disco has agreements," says Alfred.

When the boats are full the Disco and his men climb in and they push off. Then the storm clouds roll in and the rains begin. Jason and I find a boatman headed for Lebidoti. In fifteen minutes, we are on the water, speeding through the riverbones.

The water is choppy and the rain is cold. We catch up to the army, the soldiers in bright red life preservers, their commanders in sleek black rain jackets, the boatmen drenched and frantically bailing. Jason hides his camera as we overtake them. One of the soldiers gives us a thumbs-up. Then we pass a decrepit fishing boat, "Love" scrawled in blue paint on its cabin.

For the rest of the ride, Jason is silent. He shoots the lake of dead trees through the storm, his camera wrapped in plastic. The storm lasts all the way to the Ndyuka village of Lebidoti. As we come ashore, a small boy wearing blue rubber boots watches us, a garden hoe slung over his shoulder. I call out to the boy as Jason pays the boatman.

"*Mi luku Wani-pai. Yu sabi?*"

"*Ai,*" he says.

We follow the boy up a short hill past two strings of palm fronds, the same as on the hill of the dead in Balingsoela. This is the sacred gateway to the village, through which evil spirits may not pass.

The rains ease as we wander among the huts. The boy takes us past the *Gadu Oso*, the God House, a long thatched shelter ringed by the same strings of palm fronds, where the men meet to discuss community matters and make offerings to the spirits. Moments later we pass a small shrine in the middle of the footpath, again decorated with sacred fronds. A tiny girl in an immaculate baby blue dress skips down the trail, a yellow umbrella spinning over her shoulder.

The boy guides us to a well-made hut with a massive Durotank at its side, the sign of a Peace Corps home.

"*Wani-pai!*" yells the boy with the hoe.

Nathan appears at the door.

Nathan hails from Topeka, Kansas. He is twenty-six and looks as if he hasn't slept in weeks. He is six-foot-one and only 130 pounds, with dark sacs under his eyes and long, shaggy hair. Nathan is a vegetarian. He has lost thirty pounds since arriving in Suriname five months ago.

Nathan's house stinks of fresh paint. On one of the walls, he is painting a mural of the world. "For the kids," he says. On another is a collection of postcards from Kansas, fields of wheat, bales of hay, abandoned farmhouses. On his kitchen table, hundreds of photographs are splayed. Jason and I hang our hammocks and offer Nathan some Scotch.

"I shouldn't," Nathan says. "Me and booze used to be real close." He takes the cup anyway.

I tell Nathan we're here to see the goldbush.

"No problem. I can get Jamie to take us."

As we toast our arrival, thunder rolls in the distance. It gets louder and louder and then the ground begins to shake.

"Shit!" Nathan jumps up and sweeps the photos off the table into a plastic bag. "School's out."

Through Nathan's window we watch the stampede of children. They wear green and white checkered uniforms and all of them are either yelling or singing. There must be at least a hundred of them, slipping and sliding in the mud, running as fast as they can. As they pass, some of them peel away from the crowd and file through Nathan's front door. The kids pay us no notice.

"I just got these photos back today," yells Nathan above the chaos. "Word spreads fast." Nathan has spent the last few months taking photos of every child in Lebidoti. He sent the film to the city with one of the boatmen a few weeks ago.

"There's something about these kids," says Nathan. "I don't know what it is. Sometimes it really freaks me out. I've met the kids in Baku. I've seen the kids in the city. But there's something different about the children here." Nathan struggles to hand out the photos, a sea of hungry hands in his face. "And I ain't no pederast, so don't go there."

I watch the kids shoving and pushing and shouting at each other. I watch their mud-covered feet, their long limbs, the way their uniforms hang like curtains from their shoulders. I'm not sure what Nathan is talking about.

"Maybe it's the bling," says Jason.

More than half the kids are wearing a gold nugget around their necks, a gift from their fathers who work the goldbush. This is the unofficial currency of the Surinamese jungle.

The kids grow more frenzied as Nathan struggles to match each child to a photo. Soon, the bag of prints is on the ground and the kids have leapt on it. Nathan looks up to the ceiling, takes a deep breath, and explodes.

"Holy fucking Christ!" he howls, and the children stop dead. "Out! Now!" In seconds, mothers appear out of nowhere and corral their kids. Older children grab youngsters by the collars and drag them out the door. Soon the house is empty. The only evidence of the melee are the streaks of red mud across the floor.

"I'm living like a Saramaka women out here," Nathan says calmly. "I wash my own dishes, I wash my own clothes, I cook for myself. I enjoy it, but the men really let me have it. Actually, so do the women." Nathan bends down and collects the remaining photos. "But let the kids walk all over you? No way. Can't happen. You'll never survive out here if you let the kids run the show."

Nathan's friend Jamie arrives after dinner. So do a few of the children from earlier. Nathan gives the kids some money and they sprint out the door.

"Beer kids," he says. "They buy the *jugo* and deliver it cold. I give them the bottle caps. Fair exchange, I'd say."

Jamie has just returned from the city. He is not happy. He arrived on Saramaccastraat two weeks ago with 16,000 SRD of gold in his pocket. This was his salary for six months of work in the goldbush. Before he could make it to the buyers on the second floor of the Central Market, he was robbed at gunpoint. Two shots were fired. One of them went between his legs.

"*Mi libi*," he says. "*Ma mi no abi*." He is still alive but he has nothing.

The kids return with the beer and we pour Jamie a drink. Two of the children linger as we talk, nine-year-old boys who laugh giddily as they dance through Nathan's house. One of them leads and the other has his feet on top of his partner's, his arms wrapped around his waist. When they grow tired they fall on the floor and play dead.

The lights go out. Nathan strikes a match and touches it to a candle.

"That's the end of the oil," he says. "The government is supposed to send enough gas for a month, but it only lasts for a week, the fuckers. We spend the next three in darkness."

"*Yu wani si* goldbush?" asks Jamie.

"*Ai,*" I say.

"*Wi go tamara. Fruku fruku.*"

Jamie stands up to leave. The children run out the door. Then an old man appears on Nathan's stoop. He wears a faded *kamisa*, the traditional loincloth, and a moth-eaten *banya-koósu*, a cape draped over his right shoulder.

The man sits down. Nathan shakes his hand and pours him some beer. "This is Tee Delisa," Nathan says. "He stops by every night. Normally I don't give him beer. If I give him beer, he runs out and gives it to one of the pregnant ladies."

I introduce myself. Tee Delisa shakes my hand. Both his hands have six fingers instead of five. He smiles but says nothing.

"Don't worry, it's not you," says Nathan. "Tee Delisa never says a word."

I fall asleep to a distant drum.

The dreams are getting worse. I spend my days travelling this country and my nights travelling with Emma. I feel her hand on my face every time I shut my eyes and suddenly we're in the desert. The sand is flying and the man is burning and the party goes on for months. Then we're in San Marino, the smallest country in the world, where we hike the summit of Mount Titiano and ride bicycles to each other's houses. We take the train to Venice. We don masks of mud and ash and join the carnival throngs. The party ends, the piazzas fall silent, and we visit the glassworks of Murano, where we sit in the cool of an open window as swarthy Italians in tight black tunics forge purple unicorns from little else but white-hot sand.

The dreams repeat and curl back on themselves constantly, and I always end up angry. I tell Emma I need my freedom, I need to let her go. I walk deep into the jungle, try to bury her at the foot of a sacred *kankan* tree, promise to return for her when I'm ready to go home. But the tree is guarded by cats, stripes slinking through the underbrush like sex, and there is no safe place to leave her. So we do the only thing we can do. We return to Toronto, every night without fail, where we remove our clothes and explore our bodies like starved and bloodthirsty travellers.

Son-opo. Brokodei. The moon is still up and we are at the shore, *fruku fruku*, waiting for Jamie.

We sit in his boat with two other men, thick-set bruisers with heads like teed up golf balls. On either side of the dugout, children brush their teeth and playfight, spurting mouthfuls of dark water. A young girl with crossed eyes watches us nervously as she slips her dress over her head and kicks off her sandals. She holds a switch of wood in one hand and her toothbrush in the other. Both get caught in the fabric as she passes it over her head. Around her waist she wears a piece of black string, her *katitei*, the Maroon symbol of female pre-pubescence. At the earliest signs of puberty she will be given an apron; when her breasts begin to "fall to the heart" she will trade her apron for the skirts of full womanhood.

"*Yu wasi kaba?*" asks one of the men. Have you finished washing?

"*Ai,*" she says, and races up the hill with her brothers.

Jamie arrives and we push off. We motor through the bones of the bay and turn east, following an old tributary. Soon the sun appears above the canopy and the air begins to warm. On top of the tallest snag sits a huge blue parrot, the only bird I've seen on these waters, patiently drying its feathers.

An hour later Jamie pulls into a small cove. On the shore, a strip of forest has been cleared straight back into the bush. Jamie aims for this opening but before we reach land the two men in front leap out and disappear into the jungle. This is the Capasi goldbush.

Jamie tells us to wait in the boat. He walks into the shadows of a nearby shelter, where a few hammocks sag and swing with the weight of dozing men. Jamie returns with two miners. Their names are Ronnie and Richard. They have agreed to show us around.

Ronnie and Richard are Guyanese and speak good English. They have lived in Suriname since they were teenagers. Richard is in his late thirties and is a typical porknocker, very lean and extremely well muscled. He wears a FILA baseball cap and long denim shorts but his face belies his age. His stubbly beard is already greying and his skin is rough and pitted. Richard has been in and out of the interior for fifteen years and is a veteran of the Surinamese goldbush.

Ronnie looks like Richard's younger self. He wears a purple T-shirt, a pair of skimpy boxer shorts, and a turquoise skullcap. His body is also lean but lacks the brawn of a lifelong labourer. He has only been at Capasi for two months and still carries himself with the bravado of a hash dealer on Saramaccastraat.

We walk up the clearcut along the lip of the mine. On our right, the land drops off into trenches; on the left, the jungle is impenetrable. Ronnie leads the way and puffs a thick joint as he goes. Casually, he tells us he saw a jaguar up here last week.

From here we can see the extent of the operation. The Capasi goldbush is a series of individual pit mines arranged in terraces along an old stream bed, stretching about a mile back into the bush.

Each pit is a little higher than the last, a minor miracle of engineering that allows effluent and rainwater to either drain or be pumped downhill. After three days of storms, the pits have turned into ponds. Into one of them, a massive pipe spews reddish water from the pit above.

"Nobody works today," says Ronnie. "Today we pump."

"How many pits are there?" I ask.

Ronnie looks to the sky and closes his eyes to count.

"Eight," says Richard.

"And how long has the mine been here?"

Ronnie counts on his fingers.

"Six years," says Richard.

"And how did you know to start digging here?" asks Jason.

"Easy, man!" says Ronnie. "All you got to do is look!"

Until the price of gold began its meteoric rise in 2001, gold mining in Suriname's interior was mostly a subsistence pursuit. But these days, more and more Maroons leave their villages in search of jobs in Suriname's Greenstone Belt, in camps encircling the Brokopondo Lake or in remote locales like Langa Tibiki, Sela Kreek, and the Lawa River. Many mines in Sara Kreek are simply old mines that have been reopened. Erosion makes the old sites easy to find even after a century of neglect.

Soon we reach the highest terrace, the most recent pit, where the rudimentary equipment of the artisanal gold miner sits abandoned in the sun. A well-worn hydraulic monitor, three wooden sluice boxes, an ancient suction pump, and copious lengths of thick piping. We walk down from the lip of the mine and hop logs across a bog. As we peer into the pit, Richard explains the extraction process.

Workers bombard the walls of the old creek with high-pressure water from the hydraulic monitor. In theory, the resulting sludge contains high-grade alluvial gold. The slurry is then pumped out of the pit to the top of a sluice box, where it flows down a series of troughs with matted floors. Along these troughs are raised obstructions called riffles. The lighter tailings remain in suspension and stream over the riffles, eventually pouring out near the ground. But the heavier tailings – the gold – are trapped, either as nuggets against the riffles or as tiny flecks settled in the carpeted floor.

In traditional sluice box mining, the flow is stopped when the riffles are about half full with debris and the nuggets are collected. Then the riffles are removed, leaving the lucrative concentrate, or pay dirt, embedded in the matting. This pay dirt is then panned and the small flecks of gold picked out with tweezers.

But Richard makes no mention of panning.

"We stop the water," he says. "We take the nuggets. Then we pour *kwek* into the box."

Instead of panning, workers at Capasi pour mercury, or *kwek*, into the troughs. Then they rub the mercury into the pay dirt with their bare hands.

To a porknocker, mercury is liquid magic. The myriad flecks of gold bind to it to make a silvery paste. This gold-mercury amalgam is then heated with a blowtorch to burn off the mercury. Left to cool, the result is a small bead of high-grade gold.

This method is the most common gold mining technique in Suriname and among the more than fifteen million artisanal gold miners worldwide. Mercury makes harvesting the pay dirt fast and easy, a huge boon to the small-scale miner.

But the convenience of mercury amalgamation is far outweighed by its terrible health hazards. Mercury is hugely toxic. With sufficient exposure through the skin, lungs, or gastrointestinal tract, it causes irreparable damage to the human central nervous system and internal organs. Early symptoms of mercury poisoning include skin rashes and flaking, hair and nail loss, edema, sun sensitivity, and "red lips." But these visible symptoms can mask kidney and liver failure. Persistent exposure causes profound neurological damage resulting in hallucinations, memory impairment, tremors, and emotional instability. Long-term mercury poisoning is lethal, and the final stages of life are horrendously painful.

I ask Richard if he wears a mask when burning the *kwek*. He shakes his head.

"I know it's bad, but I don't have choice. Where else can I work? With two barrels of diesel, we can find two thousand American in one hole."

"Where does the *kwek* go when you're finished?" asks Jason.

Ronnie giggles. "Out," he says, pointing to the bottom of the box. Then he waves his hand toward the lake. "Away."

Every year, twenty to thirty tons of mercury are either poured into Suriname's jungles, rivers, or evaporated into its skies by artisanal gold miners struggling to make a living. Over time, the metal climbs the food chain in rising concentrations, eventually poisoning the ecosystem's top predators – the fish, crocodiles, and raptors of the Surinamese rainforest. In water with low pH and a lot of dissolved organic material – Brokopondo Lake being a prime example – the mercury rapidly converts into methylmercury, its even more poisonous cousin.

Carnivorous fish in Brokopondo Lake are the most contaminated fish in the country. The worst samples – a sad pair of piranha – registered six to eight times the maximum permissible mercury content. Fish from around here, the eastern sections of the lake near the Sara Kreek goldbush, are more poisoned than those in the west. These fish are the main source of protein for the Maroons living around the reservoir. Pregnant women and children – those most susceptible to the ravages of mercury poisoning – have little choice but to eat the toxic meat and drink the polluted water.

Unchecked mercury use is undoubtedly more dangerous to both humans and the environment than the cyanide leaching operations at the large mines. And here we have the perverse argument companies like Cambior routinely espouse in their conflicts with small-scale miners like those at Nieuw Koffiekamp: that dumping cyanide into a massive hole in the middle of a pristine jungle is more environmentally friendly than dumping mercury into rivers and streams.

At gold mines like Capasi, the majority of profit goes to the equipment owners. The labourers and the concession owner split the rest. This arrangement suits everyone because, as Richard tells us, Suriname is the only country in South America where a man can still earn his wage in gold.

"Every week, our boss pays us in gold. We hold it like this." Richard clasps his hands in front of his chest like he's praying. "Then we sell to buyers in the city or, if I am greedy, to a Chinese buyer." The streets of Paramaribo are lined with Chinese jewellers, many of whom pay more than registered buyers and smuggle the gold to Europe.

"If I like big risk," says Ronnie with a mischievous smile, "I go to Georgetown. Make twice as much."

Richard sucks his teeth. "Boy! You are too much dangerous!"

The men lead us back along the ridge of the mine. The sun is high and the heat is sticking to my skin. We stop at the camp canteen, where the workers doze in their hammocks or fiddle with their gear. A laundry line sags with sodden clothes. Two Amerindians from Mato Grosso struggle to remove a flat tire from an ancient ATV. This place is overwhelmed by sweltering exhaustion. As we bid Richard and Ronnie goodbye, an old Brazilian woman serves us shots of sweet coffee.

"You want to see women?" asks Jamie as we're climbing into his boat.

Ten minutes later we pull up to a treeless island, where we meet a Latino couple and their two beautiful daughters. The family lives here year-round in a modest shack with their flock of chickens, three pigs, and countless mangy dogs. Next to their house stands the only other building on the island, a large zinc-roofed shelter with a bar and a dance floor beneath it. The couple, in their early fifties, have owned and operated this establishment for six years. They smile proudly as we chat with their daughters. This barren island is the local brothel.

Every Friday night, Sara Kreek gold miners swarm this island to drink, dance, and purchase the affections of these two young women. Occasionally, a boatload of enterprising Saramaka prostitutes will make the trip as well; these are good nights, says Jamie. The going rate for a roll in the hay is three grams of gold, about a week's wages for a typical miner. Behind the bar is a shed the size of an outhouse. For privacy, says Jamie. It's the boudoir of Sara Kreek.

Just to survive in the interior of Suriname, the Maroons risk liver failure and brain damage in the mines. They poison themselves and their food with mercury. They smuggle gold across the border to Georgetown. But none of these activities, as challenging as they are, can possibly compare with renting your daughter to a drunk and ailing gold miner.

We return to Lebidoti at dusk. On our way to Nathan's house we pass a woman with her five small children. The youngest is wailing in her lap, while the others stand and stare at the dirt. All of them are naked and coated head to toe in a white, chalky paste.

"*Busi dresi*," says Nathan. "Jungle medicine. One of her twins is sick. The chalk is for protection."

Tee Delisa is waiting for us on Nathan's stoop. So are the beer kids. We unroll our hammocks and collapse into them.

"*Wi go sribi*," I say to no one in particular. We're going to sleep. The kids disappear. Delisa smiles and nods his head.

In the morning, Nathan takes us to Ano Kinte's house. On the way we pass Nathan's latrine, a simple shack at the top of a hill, its door padlocked. The side of the trail is peppered with piles of human feces.

"This is the crap hill," says Nathan. "Peace Corps wants me to build houses, but what they really need is an outhouse. Watch your step."

Soon we arrive at a collection of well-built huts, where lines of colourful clothes flap in the breeze. Nathan knocks on one of the doors. A tiny child opens a window shutter, a lollipop in the corner of his mouth.

"*Wani-pai!*" he yells, and Ano opens her door.

Ano is breathtakingly beautiful. She emerges from the dark of her hut like a 1940s movie star, the Rita Hayworth of Sara Creek, slipping into the late-morning sun as if it were a spotlight. Her skin is perfectly smooth, her teeth are perfectly straight, and her eyes are dark brown. She wears a low-cut red dress and her hair is gently gathered into two buns behind her ears. Around her neck she wears two chains, each weighed down with a nugget of gold.

"Hello," says Ano. "Welcome to our country. How are you?"

Ano volunteers for a local women's group, one of the growing number of grassroots organizations throughout Brokopondo aimed at Maroon women. One of Nathan's goals here is to promote women's rights with help from Ano. It is a thorny issue, especially for a white American male, as Maroon society on the whole remains fiercely patriarchal.

Ano's situation is lent further irony by her culture's rigid customs – she cannot work right now because she recently gave birth to her sixth child, and so must remain housebound for three months. She leaves her hut only to bathe in the river, twice a day. Nathan visits her every few days and the two make plans for when Ano emerges from house arrest.

Ano disappears inside and returns with her newborn.

"Emilio Gilberto Peter Kinte," she says.

Emilio is one week old and has yet to open his eyes. They are glued shut with a yellow puss that creeps down his cheeks.

"He is very sick," she says. "I have been to the doctor. He says it is the water."

I look down at the three young boys perched on the window sill. All three have that strange look to them, that eerie light in the eyes of the children of Lebidoti.

"Did this happen to the others?" I ask.

"Yes, but not this bad," she says.

Nathan and Ano speak for a while. Then we take our leave.

"Nice to meet you," says Ano. "Welcome to our country."

Beyond Ano's hut is a huge barn, which serves as the local school. As we approach we hear the excited voices and thumping feet of hundreds of students from kindergarten to grade six. Their teachers come every week from the city, living in prefabricated houses on a peninsula just past the school.

Once they have graduated from the barn, the children must go to Paramaribo to continue their education. Since most Maroons cannot afford this, their education stops at age twelve. From there, young boys are expected to work in the goldbush or as boatmen at Afobaka. Young girls are expected to become pregnant.

Across a small clearing is the doctor's house. We knock three times on his door but no one answers. Just as we're about the leave, a young Creole man stumbles to the door.

This is August, the village doctor. He was trained in Paramaribo and now runs the Poli Clinic here, a modest medical centre that services Lebidoti, Baku, and Pisheanne. We have woken him from a deep sleep. I ask him about the health of the people in the villages and he launches into a well-spoken rant.

"In the dry season," he says, "I see a lot of hypertension. The men work themselves to death looking for gold. Many of them are anemic. Many have diseases from the camps. Sexual sicknesses. Malaria. Dengue. Lechmaniosis." August points to the east. "Over there, you are taking very great risk.

"They don't work with masks or gloves. They breathe mercury. They touch mercury. It makes their nerves go badly, their brains go badly. I have seen men going crazy. Talking to themselves, hurting themselves. Their teeth go *fukup*. Their gums.

"But everything gets worse when the rains come. This is when the children begin to suffer. We have epidemics twice a year. Worms. Diarrhea. Vomiting. There is very little I can do. The people throw their trash in the bush because they have nowhere else. The people go *kaka* in the bush because they have nowhere else. And then the rains come and wash it into their drinking water." August shakes his head and rubs his eyes. "If you make latrines, kids can grow up stronger, *moro tranga, yu sabi*? If they are strong they *no go siki siki*. They can learn more."

"What about the fish?" I ask.

"The fish shouldn't be eaten. They are filled with *kwek*. Every few months a woman here gives birth to a defect. Too many fingers. Not enough toes. From the water and the fish."

I tell August about the goldbush worker I'd heard of who couldn't stop drooling. The patient's mother thought he'd been bitten by a snake, but he was actually suffering from acute mercury poisoning. And I think of Delisa's six fingers.

August yawns and slumps against the doorframe.

"It is very exhausting," he says. "The government doesn't do anything. Just me and the teachers. I see thirty new sick people every day. I don't have energy for this much longer."

I ask August what the solution is. He opens his eyes and smiles.

"Drain the lake," he says. "Bury the gold and drain the lake."

10

—

THE CONVERTING WORLD

A fobaka has returned to its licentious ways.The music is cranked, the drinking has started, the Disco has been and gone.

We find a *wagi* headed for Brownsweg. We've heard the mountain near this village offers the most spectacular view of Brokopondo Lake. Our driver's name is Bonja. Inside his *wagi*, Bonja has ripped out the benches and replaced them with plush red captain's chairs.

"Now we're talking," says Jason, sinking into his seat. "Fucking entrepreneurial."

In half an hour the *wagi* is full of northbound Maroons. Bonja cranks the music, the debut record of a young Afobakan reggae artist named Menie Mie. We listen to the tape over and over as we hammer up the Red Road. The *wagi* shakes mercilessly but our seats absorb the shocks. We doze off as Menie Mie again laments the rains that have watered down his liquor.

Two hours later the *wagi* stops rattling. We have turned off the highway and are riding on asphalt. We cross several speed bumps and the jungle on either side of the road thins out. We have reached the outskirts of Brownsweg, the largest transmigration village in Saramaka, founded by twenty-three hundred flood refugees.

I had expected it to resemble Balingsoela or Lebidoti, a traditional Maroon village. But this is a strange and spiritless jungle metropolis, a fusion of the old and new, a village of the Converting

World. We drive through the centre of town. On the left, thatched-roof huts in twos and threes, smoke from cooking fires slowly rising through the canopy, groups of children chasing each other with switches of wood. And on the right, prefabricated, government-funded houses in neat rows on bulldozed land, not a human in sight. Before the flooding, every village in Saramaka was situated on the river or one of its tributaries. Now, Brownsweg is perched on the shore of the reservoir, dependent on its polluted waters.

Bonja turns to face us.

"Do you want apartment?"

"Apartment?"

"*Ai*. Apartment."

"Yes," says Jason. "Show us to your apartments."

We pull into the only gas station in town. Behind the pumps a massive green DAF truck sits idling, bauxite mud dripping from its chassis. To the left, the Alu Bar and Restaurant. Up above, storm clouds. To the right, the edge of town, where the paving ends and the highway veers south to Atchoni.

Bonja dumps our bags on the ground.

"You live here," he says.

He points us past the pumps to the *winkel*, where a Maroon teenager stands behind a wall of rusted wire mesh, chatting with a group of grizzled men in grey coveralls. They speak Portuguese. These are Brazilian *wrokomen*, or *garimpeiros*, labourers who have come north to work Suriname's goldbush.

The boy's name is Charo.

"Apartment?" he asks.

The labourers laugh. They look like they've been drunk their whole lives. Charo comes around the back of the *winkel* and motions for us to follow him. As we walk away one of the men starts screaming and scratching his torso wildly, as if suddenly stricken with measles. His mates find this hilarious.

Charo leads us past a cage housing sick animals: three blue parrots, one baby spider monkey, a tiny white-faced saki monkey. The parrots are silent, their feathers heavy with red dust. The spider monkey cowers in the far corner as we pass, his prehensile tail wrapped around his head in fear. But the saki is curious. She climbs up the side of the cage and makes a strange peeping noise, something like a question. Her face is two half moons of cream-coloured fur, a tiny black nose in the centre.

Beyond the cage is a stand of soaring *maripa* palms. At the base of their fronds, giant seed cases hang wide open and exhausted of their fruit. Old nuts are scattered on the ground. They were once coated in the creamy flesh that monkeys adore but are now brown with rot. The Maroons grind the nuts and extrude a rich oil from the pulp, the palm oil of Suriname.

Boats have also been left to rot among the slender trunks of the *maripas*, their hulls rusted through. Around the periphery of the palm garden are six one-room cottages. These are the Brownsweg Apartments.

"Do not drink the water," says Charo as he unlocks the door to *Weekijoo*, our home for the night.

Jason and I collapse onto the cots and fall asleep quickly. Four hours later we're both awake. The rains have finished. The moon is bright. Frogs are chorusing, and a strange hum, barely audible, fills the room.

My face is suddenly itchy. Jason flicks the light on and we see them: hundreds of mosquitoes swarming near the roof. They've come inside to escape the rains. Without a word we retrieve our bug nets, but there are no hooks on the walls to hang them. There's only one thing to do: we wrap ourselves in the nets, lie back down, and wait.

"They're like fucking UFOs," says Jason.

"Hey," I say, the fabric rubbing against my lips. "Do you remember if these nets were treated with anything?"

"No," says Jason. "Why?"

The light flickers off. The humming gets louder. I feel a tap on my shoulder.

"Here," whispers Jason, glowing in the moonlight like a risen mummy. "Free one of your hands." He hands me a small plastic bottle. Whiskey.

"There's suffering and then there's suffering," says Jason. "Let's get these little shits hammered."

We talk about women, about travelling, while the mosquitoes feast.

"I can't stop dreaming about her," I say. "I'm worried she's keeping me from this place."

"The first thing I do is worry I'm going to die," he says. "What if I die? I can't die. Don't die. I always tell myself not to die."

"And?"

"That's the thing."

"What's the thing?"

"It keeps me from dying. Look, the weirdest thing about being in love is the pre-nostalgia."

"Huh?"

"Pre-nostalgia. When you feel nostalgic for a moment that hasn't passed yet. When you are so in the moment that you want it to be the only moment. And this detracts from the moment because you're worried the moment is going to pass. And so you never truly enjoy the moment."

"Pre-nostalgia."

"Yup."

We drink and scratch.

"I'm not sure that applies here," I say.

"Look. You find a woman like Emma and the world pops out." Jason takes a swig. "Sure, it's a cliché. But look at us. There's nothing more clichéd than two white dudes in the jungle wrapped in poisoned cloth, getting eaten alive, and drinking shit blended whiskey while talking about women."

"I've never heard that cliché."

"That's the thing!"

"What's the thing?"

"All I'm saying is, you're more in touch with everything when you're longing for someone."

"If you say so."

"So chill out."

We eventually fall asleep. So do the mosquitoes. They pass out, drunken *wrokomen*, on the surface of our nets.

We wake at six in the morning to the soundtrack of the Converting World. A pair of birds, caracaras, scream somewhere off in the bush. Packs of dogs howl their morning hunger. Roosters announce the sun. And from the apartment next to ours, a stereo blasts the sentimental lyrics of an old Western pop song.

Outside, one of our neighbours is smoking his first cigarette of the day and drinking his morning beer. This is Pajo, another Brazilian labourer. Pajo's eyes are clouded over and he looks about fifty. He speaks in a speedy, slurred English, repeating himself constantly.

Pajo drives a bulldozer for one of the small-scale gold miners who work the shores of the lake. He used to live in French Guiana with his wife, and like most I've spoken with here, Pajo calls French Guiana simply "France." It takes me a moment to realize he's not talking about continental Europe. For a split second, I imagine Pajo making the decision to dig for gold in the jungles of Suriname while wearing a gorgeous tailored suit and eating croissants in a café in Saint-Germain.

Pajo had heard rumours of a gold fever sweeping its way around the reservoir. When his wife died suddenly, he left his home and crossed the Marowijne River into Suriname – legally, he claims – in search of his fortune. He's been here for seven years. "Romantic gone," he keeps saying. He could be speaking about his wife or his current situation.

Since the early 1990s, an estimated thirty thousand Brazilian *garimpeiros* have come north to work in the jungles of Suriname. They've been driven here by crackdowns on illegal mining by the Brazilian government and the near-exhaustion of alluvial gold deposits in their own country. But the biggest incentive is the rumour that has been circulating the Brazilian goldfields of Itaituba and Boa Vista for more than two decades now: Suriname is the new El Dorado.

Once here, finding work is simple. Paramaribo is dotted with hotels and nightclubs that double as recruiting agencies for Surinamese concession owners. They prize the Brazilians for their toughness and resiliency, as well as their skills. Many have years of experience with such mining techniques as hydraulic monitoring, excavation, and dredging.

Many *garimpeiros* live in almost complete anonymity, working for months at a time in remote jungle camps racked by violence, malaria, dengue fever, and ecological devastation. The camps are often so well hidden that aerial surveillance can detect them only by their effluent – I've seen the photos, green-blue swirls of mercury-tinged water spilling into Suriname's rivers from jungle that looks pristine from above.

The Surinamese government is well aware the Brazilians are here. Every week, *De Ware Tijd* runs articles on the murders, thefts, and widespread contempt for the law exhibited by the *garimpeiros* and the gangs of Surinamese bandits who target them. Estimates suggest Suriname's small-scale gold industry is worth between $150 to $300 million a year, and government officials believe 80 to 90 per cent of these profits are smuggled out of the country. Yet a real crackdown has never occurred. The *mofo korantie* suggests government officials receive kickbacks from the miners in return for their silence. But Pajo doesn't know about any of this. He has never seen police at his mine site.

Pajo fumbles in his shirt pocket and pulls out a small, tightly wrapped piece of plastic. He slowly unravels it and places its contents in the palm of his hand: a twisted flake of gold, worth perhaps two hundred SRD on Saramaccastraat.

"Sometimes we find this, sometimes bigger." He wraps it back up and tucks it into his pocket.

"This is everything you have?" I ask.

"Everything. Everything. Everything."

"Where do you keep it when you sleep?" I ask.

"Up my *culo!*" he yells.

We wait for Charo's driver under a thatched-roof shelter, surrounded by young, unemployed Maroons. On the table are a half-empty bottle of Red Label and the remnants of a thick spliff. Without a word, one of the boys pours us a cup of whiskey. He waves his hand across the table, apologizing that the marijuana is all gone. Every few minutes, a northbound DAF truck pulls up to the pumps. The occupants, halfway to the city now, leap out of the cab and walk wearily to the *winkel*, where Charo sets them up with booze, cigarettes, and bags of fresh peanuts.

Our driver arrives at eleven in a rusted out pickup. We argue for a while about the price. He wants 150 SRD. We say 60 SRD. He wants

100 SRD. We say $30 US. He says $40 US. We say 80 SRD. He drives away, rolling over the tips of my boots. We call him back. We pay 100 SRD.

We're listening to Otis Redding on the eight-track and soon the paved road is nothing more than a muddy trail through thick jungle. When we reach the base of the mountain, the driver tells us to move into the flatbed. Then he floors it, tearing up the steep, mud-slicked path, earthy rainwater shooting up from the wheels. We crouch low and grip the frame as the engine chokes and the road sinks beneath us. Then the tires spin, the truck lurches backward, and Otis Redding's voice rises above the clamour.

Jason and I jump up and down above the wheels. The tires keep spinning. Our timing is off. One tire gets traction while the other loses it. The truck begins to slip sideways downhill.

The driver throws his door open. He's about to bail out. I look at Jason. We jump in unison. One, two, three. The driver gives the engine one last burst of gas. The tires catch. We swerve uphill. We keep jumping. The driver shuts his door. A half-hour later we're there.

Mazaroni Top. The highest point on the mountain. The highlight of the Brownsberg Nature Reserve and arguably the best view in the country. Far below and stretching to the horizon, the lake shimmers between cloud shadows, the riverbones a carpet of brown on its surface.

Jason sets up his tripod and starts shooting. We spend the afternoon smoking Morellos and drinking what's left of the booze. A massive storm rolls toward us and then veers south. Rainbows arch from the water and land in invisible places. For the first four hours we are completely alone. Then we hear laughter from the bush below. Two men emerge at the trailhead. They climb up to the lookout and slump beneath the palm-frond shelter.

The two are Dutchmen, Bryan van Dyck and his younger brother, Olly. Bryan is stocky and has a worn, sunburned face. Olly is taller and better built, with a heavier brow and a shaved head. Olly is visiting from the Netherlands.

"This is beautiful place," says Bryan. "It is like oasis."

Bryan owns the biggest pig farm in Suriname. He runs his business from a large piece of land near the Saramacca River and comes here once a month. "Here I don't have to think about life," he says, "or my animals getting sick. Or all the Jews buying cows. I just *koiri*."

"Do people hunt in Canada?" asks Olly.

"Yeah. Lots of people hunt," I say.

"What do they shoot?"

"Deer. Moose. Bear."

"You have many bears?"

"Yeah."

"What kind of gun do you use for bear?" says Olly.

"I don't know."

"Come to my farm," says Bryan. "I will set you up with an ATV. We will hunt wild boar."

The sun drops down. The trailing storm clouds turn pink then orange as a cool breeze picks up. Far to the north, at the top of the lake, the lights of the Afobaka Dam blink on. Then, from somewhere down below comes a gruff bark, like a dog trying to clear its throat.

Olly leaps from his seat. "What the hell is that?"

Bryan bursts out laughing, yells something in Dutch. "My big, tough brother!" he says in translation. "It is only baboon!"

The barks get louder, more frequent, and soon they join in a haunting, melancholic moan. We cover our ears. These are the calls of the *babun*, the red howler monkey, the loudest land animal on earth. They are no more than ten metres beneath us, hidden in the canopy, reminding nearby troops that Mazaroni Top is their territory.

"What do they want?" asks Olly.

"Nothing," laughs Bryan.

"Then why do they scream?"

"They aren't screaming," says Bryan. "They are talking."

After five minutes the chorus stops, and the first monkey, the alpha, gives one last, aggressive grunt. Olly sits back down, fingers the straps on his pack nervously. Bryan lights another smoke and mocks his brother mercilessly.

A thick mist has rolled in from the lake, shrouding the lookout and eliminating the view. I look behind us and confirm the rumour – the lights of the Suralco plant, far below and to the north, glimmer through the gloom.

"One of our uncles owns an island over there." Bryan points into the fog. "It doesn't have a name. But he takes tourists, shows them the jungle."

"We call it Rich People Island," says Olly. "He takes you to a village on the mainland where those people live. What you call them? Not the Indians, the other ones. The ones that look like apes."

"Maroons," says Bryan.

"Maroons! He takes you to see Maroons, to learn how they live. You can take photographs, buy calabash, walk in the forest, relax in your hammocks. It is very much cheap. Very good thing for Suriname, my uncle."

I begin to feel anxious and annoyed, the same way I felt back in Raleighvallen just before I left the reserve. I sense a connection, however tenuous, between Conservation International's efforts in the CSNR and this man who runs Rich People Island.

Ecotourism's practitioners preach harmony with the natural world as they escort us into the planet's most fragile, pristine places. Couched in the soothing rationalism of science, it has achieved a remarkable cachet. Costa Rica used to cut more of its forest than any other nation on earth; now tourism dwarfs the timber industry there. These days, ecotourism is the gospel of modern conservation.

The problem, though, is that ecotourism encourages the very privileged and arrogant behaviours that lie at the root of our planet's ecological crisis. Charter a plane to the other side of the globe! Be the first to set foot in a pristine rainforest! Ride a zip-line through the canopy of a cloud forest! Collapse into your hammock, cocktail in hand, and drift off to the haunting voice of the howler monkey! True, it has positive qualities, such as an economic engine not based on extracting resources, and the distribution (in theory) of monies to poorer regions of the globe. But to call ecotourism a solution, instead of merely a stop-gap, seems increasingly strange to me. Listening to Bryan and Olly, it occurs to me a real solution to our environmental woes lies somewhere deeper than mere business economics, somewhere beyond the "existence value" of a swath of trees.

"You should call me," says Bryan. "We will go hunting in the lake."

"I like to fish," says Jason.

"Not fishing. Hunting. On the islands. It is very much fun."

"There are animals on the islands?" I ask.

"Of course!" says Olly, standing up to leave. "The big ones were restocked with deer and pigs."

Bryan hands me a slip of paper with his cell number on it. "It is beautiful place," he says. "We go once a year. The dogs scare the animals out of the jungle and into the water. Then we shoot them."

Olly smiles and waves goodbye.

"It really is very much nice."

We walk back to our hammocks by flashlight. The night is cold, the mountain shrouded in mist. We wake to the howlers a couple of times, their cries carrying from somewhere south of Brokopondo.

Contradictions are a defining feature of the Converting World. Whether it is a Maroon man helping raise a wall of cement that will eventually drown the village where he lives, government officials allowing Canadian companies and thousands of *garimpeiros* to plunder Suriname's riches, or simply animals being returned to the islands their predecessors were rescued from forty years ago just so they can be slaughtered, the people who live in this world live at a vague and troubling crossroads – between right and wrong, between sustainability and devastation, between hope and memory, between survival and extinction. Everyone here, just like everyone everywhere, is struggling for a leg up, for some kind of traction up the mountain, to convert what they've already got into what they desperately need. And what's worse, they are subject to the judgment of people like me: people from a society that is convinced it has moved beyond this divide – that it has "developed" and, therefore, has economic authority over those who are "developing."

The worst part about listening to the van Dycks was not their contradictory attitudes or even their racism. It was realizing their contradictions are the same as mine. After all, Canada allows animals to be hunted for sport. Canada spews mercury into the air from coal-fired power plants. Canada allows shipping corporations to avoid taxes by flying the flags of Caribbean nations. Canada, multicultural icon of the world, still allows many of its First Peoples to live in extreme poverty on reserves without proper sewage systems or clean drinking water.

Canada is plundering Suriname's gold.

While it is easy to judge others in matters of conservation and government and label them hypocrites, we should ask how the West managed to convert so quickly and so effectively. The answer is no secret: it all happened at the expense of others. Much as we try to convince ourselves otherwise, we are no better than the Maroon who floods his village and poisons his children with mercury for a wage in gold. No worse and no better.

The lesson I've learned is that labels like "converting," or "developing" are equally misleading and cause more problems than they solve. Because we all live at that same crossroads between right and wrong, between short-term gain and long-term survival, between happiness and heartbreak, no matter the state of our economy, or society. This is not the poor person's dilemma; this is the human dilemma. We all live in the same state of longing.

In the morning, a government *wagi* takes us down the mountain free of charge. Near the slash-and-burn gardens on the outskirts of Brownsweg, a young man walks bent over, a rusted-out engine on his head.

11

METAMORPHOSIS

It is tempting to believe the only thing that doesn't change here is the jungle, that the forest is the same as it's always been, a constant flourish of green and brown on the periphery of this lake. But this misses the point of the jungle entirely.

Three hundred years ago, the Dutch naturalist painter Maria Sibylla Merian knew this better than most. In 1699, at the age of fifty-two, she travelled to Suriname to study and paint the insects of the New World. She lived here for two years and spent much of her time searching out and painting her favourite subjects – the caterpillars, cocoons, and chrysalises of moths and butterflies – on the numerous sugar plantations that lined Suriname's major rivers. Merian was captivated by metamorphosis, the process by which caterpillars become winged insects, and made it the central theme of her work.

Four years after being driven from Suriname by malaria, Merian published her masterpiece, *Metamorphosis of the Insects of Suriname*, a folio of sixty copperplates comprising a beauty and scholarship never before seen in naturalist painting. Her vellums portrayed the entire life cycles of her subjects, from caterpillar through chrysalis to butterfly, and at the centre of each portrait were the stunning plants that served as the insects' natural foods. *Metamorphosis* was lauded as "the most beautiful work ever painted in America" and Merian quickly became something of a celebrity artist in Europe.

In Merian's time, a revolution in classification was taking place in the biological sciences. This was around the time of Carolus Linnaeus, the father of modern taxonomy, when wealthy men fancying themselves as naturalists mounted expeditions to the far reaches of the globe and returned with astounding collections of previously unknown plants, animals, and insects. These were the earliest days of modern scientific imperialism, and the new life forms were quickly given Latin names and incorporated into a soothing taxonomy, the same system of nomenclature biologists use today.

In this context, Merian's work takes on a subversive flavour. While the men of science were busy cataloguing the diversity of life, Merian focused on the interactions between life forms – the caterpillars, plants, spiders, bees, and birds that make up the whole of an insect's world. For Merian, the only way to understand how caterpillars wrapped in silken pouches could possibly emerge, months later, with gorgeous wings was to paint them *in situ*, not removed from the wilderness but within the intricate contexts of their lives. Merian portrayed in her paintings exactly what she had witnessed – the utter interconnectedness of the jungle.

Metamorphosis of the Insects of Suriname remains a masterwork of naturalist art and an early triumph for what we would now call the ecological way of thought. It also emphasizes the sole truth of the rainforest: that although the jungle seems permanent, standing as it does like a brooding giant on the outskirts of my life here, the forest is constantly transforming from one mode of life to another.

Every day, the storms build. The Brazilians sweep north, the mercury accumulates, and the Maroons struggle for rights to life. The children grow sick, the lake grows shallow, the water grows more and more fetid. The economy falters, the gold mine leaks cyanide, and the people expect disaster. And the forest continues to eat and give

birth to itself, dirty moths to gorgeous butterflies, decay transforming to life.

It's almost Christmas. Tomorrow we are leaving for the east and
won't be back until New Year's Eve. I call Emma to wish her a happy
holiday.

She is not doing well. Her fears of abandonment are sneaking
up on her again. She screams – *Where are you? Where the fuck are
you?* – but I can't give her an answer. *I'm trying to figure out where
I am. That's the fucking point.* But this only makes her angrier.

I tell her about Fritz von Troon, how he's going to help me find
okopipi, how I just have to bide my time until the New Year. I tell
her about the December Murders, about Bodi, the truth he refuses
to share with me, the truth I'm desperate to hear. Then I tell her
about my favourite word in Sranantongo, *koiri*, the wandering
without a purpose, the act of faith this word implies.

Emma is in no mood for acts of faith. She says she is weaker
than me, that these are my ideals, not hers. Then she says something worse, something I've always hated about myself, something
I thought she loved in me but apparently not. I ask where this is
going. She tells me nowhere good.

Go find what you're looking for. Go find your stupid blue frog.

We mourn in silence on the phone. We are at the foot of that
kankan tree, shovels in hand, surrounded by cursed jungle. I listen
to her weep as if her head were on my shoulder; she does the same
for me. Then she gets ahold of herself and I hear her swallow.

"You've met someone new," I say.

"Maybe," she says.

My heart snaps for the very first time.

PART TWO

Borderlands

"O Pangloss!" exclaimed Candide, "you had not guessed this abomination; this does it, at last I shall have to renounce your optimism."

"What is optimism?" said Cacambo.

"Alas," said Candide, "it is the mania of maintaining that all is well when we are miserable!" And he shed tears as he looked at his Negro, and he entered Surinam weeping.

VOLTAIRE, CANDIDE

12

—

THE HOLIDAY TOWN OF ALBINA

The sweet woman at the ticket booth is taking her sweet time. She shuffles papers behind her cage as if it were a lazy Sunday, as if she'd been called into work on her day off. In the parking lot the Albina bus is ready to go – the passengers are sweating, the engine is idling, and the driver is on the roof, wrestling the last of the cargo into place. Everything has changed in this country in the space of one day. This morning, the buses in Suriname are leaving on time.

The rains begin. Jason runs to a nearby *winkel* to buy garbage bags. Finally, the woman hands me our tickets and I drag our packs outside. We wrap them in plastic and heave them up to the driver, who tells us there is no room. We insist and show him our tickets. He swears at us, loosens the ropes, and shoves our gear beneath.

The passengers aren't happy, either. Though they try to stop us, we squeeze through and find seats at the back. I wedge myself between an enormous woman who tells me her name is Glory and an old Ndyuka man who is already fast asleep on the shoulder of an old Ndyuka woman. Jason is one row in front, wedged between two arguing brothers. The men yell at each other while Jason cowers between them like an outgunned marriage counsellor.

We take Saramaccastraat south to the bridge, the long balustrade of concrete that sweeps across the Suriname River, the one Jules Wijdenbosch opened to great fanfare in 2000, two days before he

lost the election. "Never trust the political heart of a Surinamer," a friend once told me. The whole country turned up for that ribbon cutting, for the all-day party, and then handed Venetiaan a landslide victory.

From the height of the bridge, the strategy of Paramaribo's founders becomes clear; the city is built at a sharp bend in the river, sheltered from the sea. From Fort Zeelandia, on the elbow of this curve, military commanders would have had perfect sightlines up and down the river.

From here, Paramaribo looks like a sleepy town, a postcard village by the water. Far below, an armada of dugouts plies the water, ferrying people back and forth between the city and the District of Commewijne. They pass on either side of the *Goslar*, the black wreck of a German war boat that was scuttled during the Second World War. The crew had been taken into custody by the Dutch, but the captain was allowed back on board to collect his clothing. According to plan, the ship was anchored in the deepest trench of the river and its entire coal ballast had been moved to one side. The captain simply opened all the portholes on that side, sinking the ship and hindering for the duration of the war shipments of Surinamese bauxite, the main source of wartime aluminum for the U.S. Air Force.

We join the east–west highway. Roadside towns give way to road-side villages, and then the jungle begins. We are headed east, toward the old stronghold of Ronnie Brunswijk and his Jungle Commando, a borderland that still seethes with an anti-government, pro-Maroon power politics. Deep into the Marowijne region, where Petrus is from, where the memorial for his mother will take place in four days.

This is the most dangerous stretch of road in Suriname. The farther into the bush you go, the more likely you are to be attacked by bandits. They may be members of a drug cartel or a local gang or simply amateur highway robbers. The road is in terrible shape, the asphalt riddled in places with cracks like fault lines: perfect conditions for a roadside ambush.

The rain crashes down, the stereo is on full blast and my mind digs in for the long haul. Two months of dreams come rushing back to me and so do Emma's words from yesterday. I lean my head against the seat in front and silently scream at myself – not here, don't do this here.

The driver slams on the brakes. The passengers shriek and Glory leaps out of her seat. This must be it. We've reached a roadblock and are about to be set upon by a mob of bandits. For a moment, my gloom is overwhelmed by a more visceral despair. Jason fumbles for his camera, people lay hands on other people's shoulders and the bus fills with frightened talk.

There are no bandits. There's been a horrible car crash. Two small sedans – their windshields smashed, their hoods crumpled, their frames charred black – face each other in the left-hand lane. They look like two old warriors staring each other down for one last charge, manoeuvring for the final blow. Our bus inches past and the passengers fall silent. No one could have survived this accident.

Then I see the sign for Moiwana and am embarrassed by my mood.

On November 29, 1986, at the height of the civil war, the Surinamese National Army attacked the Ndyuka village of Moiwana, killing thirty-nine unarmed men, women, and children and burning their village to the ground. The unprovoked attack was the most horrific in the nation's young history. Witnesses reported seeing babies shot point-blank while sleeping in their mothers' arms, groups of pregnant women strafed with gunfire, and men hacked to pieces with machetes. Seven out of every ten victims were below the age of eighteen; one in four was younger than five. Those who escaped fled deep into the jungle. It's rumoured one man saved himself by hiding beneath an outhouse. For two days he was up to his neck in feces, too afraid to move.

A third of the survivors made it to Albina and the Marowijne River, a three-day trek through the bush. They crossed over to French Guiana and joined the thousands of Maroons who had already fled the war and sought shelter in refugee camps set up by the French government. The rest fled either to Paramaribo or to Moengo, the large mining town run by Suralco and BHP Billiton. None of Moiwana's dead were properly buried. Some were dumped in a mass grave while others were cremated. Maroons consider cremation a repugnant act sure to anger the spirits and, as one witness put it, "burden all the children."

Days after the attack, Army Commander Desi Bouterse announced publicly that he had planned and ordered the massacre. He was simply following through on a warning he'd issued in a radio address weeks earlier, that he would kill all Maroons.

A group of survivors soon formed Moiwana '86, an organization tasked with bringing the perpetrators to justice. Although Bouterse refused to allow the Surinamese police to investigate the attack, the man in charge of the state's criminal investigations, Inspector Herman Gooding, made brave headway in identifying the perpetrators. To put an end to his inquiry and intimidate his colleagues, Gooding was dragged from his car in front of the Zeelandia military barracks and shot to death. Gooding's assassination has never been investigated.

Along with the December Murders, Moiwana is Suriname's other enduring trauma. The massacre became synonymous with Bouterse's policy of visiting reprisals on innocent civilians for the wartime activities of the Jungle Commando. Nineteen years after the bloodbath, it continues to haunt this stretch of the east–west highway. The old village is still a ghost town and the Surinamese government continues to refuse requests from Moiwana '86 to open a full investigation.

But soon it will be forced to. In June 2005, the Inter-American Court of Human Rights convicted Bouterse's military regime of violating at least five articles of the American Convention for Human Rights on that day in Moiwana. The Court demanded reparations be paid to the survivors, the bodies of the dead be repatriated, the legal title of Moiwana lands be handed over to the Ndyuka, a public apology be made by the Venetiaan government, and a full-scale investigation be launched to prosecute the "intellectual authors" of the massacre "for crimes against humanity." Three weeks ago, a public memorial was held here by the survivors and relatives of the dead, the first time they'd been back to their ravaged home in nineteen years.

Ronnie Brunswijk, the former leader of the Jungle Commando and current member of parliament, arrived at the memorial by helicopter to give a speech. He gave thanks for the Court's ruling, especially the survivors' reparations of nearly US$3 million (each survivor eventually received approximately US$13,000). But Brunswijk also vowed, "This will not take away the pain in our hearts." Until those responsible for the massacre are rigorously punished, no amount of money will ease the anger of the dead.

An hour later, our bus pulls into the Albina market, where it is promptly set upon by a throng of men. They bash on the windows and scuffle with each other for position. All of them point to the east.

"*Tap'sei?*" they yell. "*Tap'sei?*" They want to take us "topside," across the Marowijne River to the town of St. Laurent in French Guiana. These are the ferrymen to France.

Our driver climbs up to the roof and unties the gear. As we wait to catch our packs the boatmen swarm.

"*Tap'sei? Tap'sei?*"

"No *tap'sei*," I tell them. "*Wi go Nason.*"

"No France?" they ask.

"No. No France."

Now a teenaged Carib boy approaches. He wears a ponytail, a white headband, and warm-up pants with side buttons. He is a smooth-looking boy but for his enormous overbite. He could be the Surinamese Ronaldinho.

"*Yu wani* hotel?" he asks.

"*Ai.* Hotel."

"Come." He shuffles away from the crowd. We shoulder our packs and follow.

He takes us past shanty houses, where naked Carib and Ndyuka children play together in flooded front yards, the village cemetery spilling over with crumbling headstones, a wrought iron gateway guarded by three Surinamese soldiers. Behind these soldiers sits a half-demolished building, piles of scorched rubble among slim trees. Before the interior war, Albina was a popular holiday town where young families from Paramaribo could enjoy a relaxing day by the water or take a day trip to France. Then during the civil war, the army set the town on fire and burned much of it to the ground.

The hotel is deserted. The rooms are locked and no one appears when we ring the service bell. The walls of the lobby are brown with mould. The only sign of life is a wide-screen television perched on the front desk. David Lynch could film a feature here.

Our guide looks confused. He closes his eyes to think, and an explosion splits the air.

"Business not so good," he says.

We walk back the way we came.

The Albina Breeze Guesthouse is located on the second floor of a well-kept building just west of the market. On the ground floor is the only pharmacy in town, where pregnant women come to buy Epsom salts, where teenaged boys come to buy diapers. A Javanese man named Mike owns both businesses and agrees to rent us a room. He also knows a boatman named Mr. Jack who might take us upriver.

In town, at the Chinatown Supermarket Bar and Restaurant, we eat enormous helpings of smoked chicken, breaded shrimp, cubed ham, and fried noodles while a movie set in ancient China plays on a black and white television. We watch Genghis Khan–types butcher each other with gleaming scimitars as beautiful women in long, white robes stroll through steaming battlegrounds minutes after the war is won. Our chef doesn't speak Sranantongo and we don't speak Mandarin. He scribbles our bill on a napkin and then falls asleep behind the counter.

At the water's edge, the constant buzz of outboard motors. Women washing dishes in the shallows, men pissing among the rocks, kids smoking hash in the ruins of old lookout posts. A Hindustani family of three arrives by boat from France. A Chinese family of eight takes their place and is whisked off to Europe.

The shore is strewn with plastic packaging and glass bottles and wads of newspaper. For more than a month, Albina's garbage has been piling up in the streets. The government is behind in paying trash collectors, so the rubbish just sits and rots. Last week, the people of Albina banded together and spent five days burning as much garbage as they could. The event was called *Krin Albina*, and posters hung on every available lamppost and tree; the effort was a modest success. But now rainstorms have ripped the posters down and added them to the litter.

Behind us, soldiers labour behind a fence. They push wheelbarrows piled with blackened rubble. Jason pulls out his camera and a man nearby hollers at him. Do not take photos of the army, he says. Soldiers do not like photos.

Dusk falls and we walk back to the Albina Breeze. Another explosion booms from the market and echoes back from the French jungle. At the guesthouse, we wait for Mike and Mr. Jack in the upstairs kitchen while a prostitute and her john have vigorous sex in the shower down the hall. The woman sounds exhausted but happy; the man moans like he's having a strange dream.

Mr. Jack is a Ndyuka Maroon who lives with his wife in a ramshackle house across the street. He is fifty years old with a barrel chest, stumpy legs, and a very serious demeanour. He has been a boatman here for more than thirty years, speaks very little English, and readily admits to a near-surface hatred of tourists. He and Mike have known each other since Mike was a child.

Mike is ex-army. He is short and carries himself with the calm confidence of a black belt in the martial arts. Every now and then he slams the palm of his hand against his jaw, two times in rapid succession, cracking his neck left and right.

Mr. Jack launches into a description of his talents. If we are hungry, he will catch a fish and cook for us. If we are thirsty, he will catch the rain. If we need light, he will bring us lanterns. If we get bitten by a snake, he will suck the poison from our flesh. If we need a woman, he will find us one. Then he lists the places he can take us. For 6,000 SRD we can go to Stoelman's Island, the old headquarters of the Jungle Commando. For 8,000 SRD, we can go farther and meet the chief of the Aluku Maroons near the border with Brazil. There are many hammock camps upriver, he says. For the right price, he can take us anywhere.

I tell Mr. Jack we do not have much money. "We want to go to Bigiston and then Nason," I say. "There is a celebration for my friend's mother on Boxing Day. I want to pay my respects at her grave."

Mike translates and Mr. Jack nods. "No tourist," he says.

"Not really," I say.

"Thank you," says Mr. Jack, the only English he knows.

We settle on a price, almost everything we've got. Then another explosion rings out from the market.

"It is Christmas," says Mike. "From now until New Year's Eve, we set off fireworks to celebrate."

I ask Mr. Jack where he was born but he doesn't know. It doesn't matter, he says, because the Ndyuka are born and raised on both sides of the Marowijne River and consider themselves first and foremost members of the Ndyuka nation. But he is also a citizen of France and Suriname. He has large families in both countries.

He asks me where my home is. I tell him Raleighvallen.

"You live with Kwinti?" he asks.

"Ai."

"The Kwinti are not to be trusted."

"Why not?"

"Because they are almost gone."

Mr. Jack turns to Mike and whispers something.

"He wants to know if he should trust you," says Mike. "The last time he went upriver he took four Dutchmen to Langa Tibiki. They were drunk the whole time and embarrassed themselves. Now Mr. Jack can't go back there."

I quote my taxi driver in the city, another man who dislikes tourists. *"Tumsi jugo jugo no bun,"* I say. Too much mischief is no good.

Mr. Jack smiles and stands up from his chair. "A *bun*," he says, reaching out his fist. I thump knuckles with him and so does Jason. I have spoken a secret Surinamese code. We have a deal.

We buy two *jugos* of beer and try to drink ourselves to sleep. The ceiling fan whines and the fireworks get closer and the rain explodes on the roof. Jason thinks it's funny that I consider Raleighvallen my home. I tell him this is a recent development.

Like clockwork, with the booze comes the sadness. Jason leaves the room when he sees I'm about to lose it. The loneliness that's been threatening since I arrived in this country drops down around me. I try to sleep but my mind has other ideas.

None of the Moiwana survivors have returned to their old village. Many are still in Paramaribo or Moengo and more than one hundred still live in French Guiana. This is because those responsible for the crime, especially Desi Bouterse, still wield enormous political power here. Intimidation and coercion are alive and well in the Surinamese parliament. Although it's been nearly two decades since the massacre, many Ndyuka still fear reprisals should they return.

Beyond the physical suffering, the survivors of Moiwana are spiritually tormented. As one witness told the Inter-American Court, "It is as if we are dying for the second time." The Court's judgment states the issue precisely: "Not only must the Moiwana community members endure the indignation and shame of having been abandoned by Suriname's criminal justice system – despite the grave actions perpetrated on their village – they must also suffer the wrath of those deceased family members who were unjustly killed during the attack."

Although the government has pledged to rebuild the village and hand the Ndyuka legal ownership of the land, it is hard to imagine the survivors ever returning to Moiwana. Unable to fulfill their obligations to the dead, the Ndyuka feel, as another witness put it, "as if we do not exist on earth." Ruthlessly uprooted and forced into exile, their hearts are profoundly broken.

This morning, the lesson of the jungle is this: there are startling moments of beauty in some of the darkest places. They catch your eye even as you stare blankly at everything.

There is a young papaya tree behind the guesthouse. It reaches through the piles of garbage and rusted sheet metal and hacked banana plants, its trunk still slim and pithy but soon to be woody and scarred. Its spiralling leaves are more green than anything I've seen in this country. And here is where the beauty lies. Rainwater has collected at the heart of each leaf. These small pools reflect the sun and have turned into tear-shaped beads of silver.

The termites have hatched in Albina. We wander through the market, buying booze and groceries while the air writhes black around us. In darkened stalls, young boys sit still as young girls carve tattoos into their shoulders. An old Ndyuka woman sells oranges and bananas and throws in an extra bunch to thank us for our business.

Open umbrellas lie discarded in the dirt as if their owners had vanished beneath them. Kids throw lit fireworks at our feet and scream as we hop the blasts.

We meet Mr. Jack at the south end of the boat launch. His boat is long and beautifully straight, with sheets of metal nailed along the gunwales. He has bought two cases of Parbo and stashed them in the bow.

Mr. Jack pulls the engine to life and points to the sky. God will bless us if we reward those who help us, he says. Our trip suitably sanctified, his tip ensured, we begin our journey upriver.

13

—

A NOISE MADE LONG AGO

South of Albina, the riverbank is a graveyard of failed and forgotten projects. The rusted hull of a freighter lies half out of the water, covered in creeping vines. On a small hill sits a half-finished mansion, which was going to be Suriname's biggest nightclub until the owner suddenly died. Nearby is the skeleton of a black Mercedes Benz, its interior pillaged, its doors long gone.

We pass the settlements of Papatam, Martin, and Akoloi Kondre, where Mr. Jack yells to the women on the shore, two of them his sisters. We cross to the French side, past Pampilla Island, to the village of Potale. This used to be a white man's village but now it is all Ndyuka. The water here is dark green, transitioning from the blue of the sea to the brown of the bush. Mr. Jack tells us the name of every creek and every fruit tree we pass.

A half-hour later, we pull to shore beside an enormous, half-submerged rock. The stone is shaped like a dome and is encircled by petroglyphs, stick figures of women and men carved into the surface. Next to this ancient artwork, a message is scrawled in white paint: *Bigiston, Suriname, and the Earth are Beautiful. Please, SVP, No Dirty in the River.*

Mr. Jack says he'll be back for us early in the morning. We unload our gear and hike up to a group of huts with thatch walls and circular roofs, Amerindian style. An old Carib man wields a

chainsaw beneath a butchered mango tree. Someone yells my name.
My friend Rupe is home.

The Marowijne Region of Suriname has long been the land of the
Ndyuka and Paramaka Maroons. But before the Maroons arrived,
this area was Carib territory. A few Caribs still live here, in small
villages between Albina and the sea, and Bigiston is the southernmost village with a sizeable Amerindian population.

Bigiston is a composite of four distinct villages spread along the
river – the Ndyuka quarters of Damba Kondre and Kronto Kondre
and the Carib quarters of Tamarin Kondre and Bigi Kondre. Rupe,
another of my Peace Corps friends, lives in Tamarin Kondre.

Rupe takes us on a short *koiri*. In the slim patch of forest that
divides his neighbourhood from Kronto Kondre is a small clearing
filled with Maroon dugouts at various stages of construction. Some
of them are still black with fire, their hulls having recently been
burnt out, thick slabs of purple-heart wood jammed between the
gunwales to keep their shape as they cool. Others are nearly finished,
their bows already painted with reds, blues, yellows, and whites.

In Kronto Kondre, the houses are clearly Maroon, placed close
together with milled siding and zinc roofs. But the huts of Bigi
Kondre are of *ingi* (Indian) design and decoration: the homes have
more thatch, often for the walls, and are more spaced out. Inside,
gorgeous white cotton hammocks hang next to woven *matapis*, the
long, tubular cassava presses used to make *casiri* beer. Pinned to
one exterior wall is the rotten carcass of a harpy eagle, the most fear-
some bird of prey on earth, the species that used to stalk my monkeys
for months on end. The bird's wingspan is at least two metres, its
talons four inches long and razor-sharp.

Rupe has been here for a year and a half and hails from Los

Angeles. I ask him what it's like living with Ndyuka and Caribs at the same time.

"It can be awkward. Before the war, Bigiston was 80 per cent Carib. Now it's 60 per cent Maroon. The Ndyuka consider this their village because they did much of the rebuilding. They don't recognize the *ingi* captains. Whenever there's a big decision to make, things get heated."

Rupe has one more thing to show us. He takes us to the village school, in a clearing overlooking the river. At the south end, a lone concrete wall is swamped with creeping vegetation and stained black with ash. This is the last remaining wall of the old school, burned to the ground during the hinterland war. When I get close I notice the bullet holes riddling its surface.

We end our *koiri* with a swim around Bigi Stone. The petroglyphs are mostly underwater and very faint, as if the carver had second thoughts about defacing such a gorgeous rock. Up close, the figures lose their structure and become little more than smooth depressions in the rock face. I run my fingers over them and try to imagine their meaning.

Rupe has brought his fishing line and some flour for bait, and soon we are fishing our swimming hole for dinner. It takes an hour, but Jason finally pulls something from the water – a small, reddish piranha. With its flesh as bait we quickly pull in more of the creatures, their jagged teeth spilling out of their mouths, their bodies rigid and tank-like. Our swimming hole is teeming with them. Each fish is bigger than the last. We walk back to Rupe's with a string of five as Jason plans the marinade.

Seyfu is waiting for us on the porch, sitting in front of a mouldy checkerboard and smoking a hand-rolled cigarette. He has heard about the white men who have come to learn about his home. He has come to tell us stories.

Seyfu is a Carib of perhaps fifty. He has a distinguished, handsome face with a perfect long nose and lips turned up into a soft smile. He carries himself with the calm reserve of a gentleman, a jungle aristocrat in a light green jacket, a torn red sweater, and blue plaid pyjama pants.

There is a hole under Bigi Stone filled with gold, he says. It is the home of a white man, a white woman, and their children. A black slave lives there as well. Without a hint of irony, Seyfu calls this slave the Boss. Legend has it that if your wife swims too close to the stone, the Boss will steal her from you.

One day not long ago, two Dutchmen arrived in Bigiston and dove beneath the rock. They found the storehouse of gold and brought it to the surface. Then they proceeded to argue. One of them wanted to take the gold home to Europe; the other wanted to leave it where it was. This man eventually lost the argument and soon the gold was in Albina, loaded in the back of their truck.

This angered the Boss, says Seyfu. The Dutchmen were guilty of *takru prakseri*, or evil thoughts, so the Boss put a curse on the men. The next day, on the road to Paramaribo, the man who had wanted to steal the gold inexplicably died at the wheel. His partner, who understood the power of a Maroon curse, quickly turned the truck around and returned the gold.

Seyfu slowly drags a red checker across the mouldy board. The Bigi Stone is always moving east, he says. It yearns for France, for Europe, and little by little it is trying to escape there. But whenever someone looks at the stone, it returns to the shore. As long as the people of Bigiston keep an eye on it, says Seyfu, the rock and the riches beneath it will remain where they are.

The legend of Bigiston encapsulates the whole history of the South American continent since the Spanish Conquest began.

There is gold beneath our feet; if we do not pay attention, someone will come and take it away.

Seyfu asks for paper and a pen and tells another story as he draws. There is a man in the jungle who is good, he says. His name is Kalupi. He makes a lot of noise, banging on the buttress roots of trees, yelling and singing ancient songs. Kalupi will help you when you are hungry. When you hear his voice, you will fish and hunt with success. But there is another man, says Seyfu, an evil man named Tekokuh. He has long hair and sounds like the wind. When he screams you cannot walk and you cannot hunt. It is like trying to hike into a hurricane. Tekokuh eats escargots by the hundreds. The only sign of him are the piles of snail shells scattered throughout the bush.

I peer over Seyfu's shoulder. He is drawing the petroglyphs of Bigiston. Although many think they are pictures, Seyfu tells us most of the carvings are letters. They spell three words from an ancient Cariban language.

The first word, *peroe-roetepo*, means "Pururu tree on top." Seyfu says this refers to the slim trees growing on top of the stone. The second word, *petapotepo*, is a mystery to him. He's asked all the elders of Bigiston but none of them know what it means. The third word, *etapotepo*, means "a noise made a long time ago." Seyfu is not sure, but he thinks it has something to do with the war.

I ask if *etapotepo* might mean you can stand on the Bigi Stone and hear voices echoing from upriver, voices from the past.

Seyfu smiles. "*Ai*," he says. "*Kande*."

The rains return during the night. I lie awake listening to the storm. Then the rope on my cheap plastic hammock gives way and I crash to the floor. I stumble through the dark into Rupe's room and climb into the cotton hammock his neighbour Wilhelmina made him. I am instantly enveloped in a cozy cocoon. The Caribbean is the home of the hammock, a word rooted in the Arawak term for fish net.

In the morning, every bucket in Tamarin Kondre is overflowing and the river has risen almost a metre. *Ingi* mothers braid their daughters' hair for the coming Christmas celebrations while *ingi* fathers check their fishing nets. A lone canoe is mired in the reeds downshore. This is an abundant time in Suriname. The short rainy season is nearing its height; the gardens, the trees, and the people are drinking their fill.

While we pack our gear and wait for Mr. Jack, Seyfu appears again. He sits down, reaches for a cigarette, and crumbles the tobacco onto a small piece of fragile brown bark. "*Ingi pipa*," he says. Carib rolling paper. Seyfu skilfully rolls the cigarette, wraps another thin piece of bark around it, and ties a bow. He lights it and shares it with us while he scribbles again on the same piece of paper as yesterday. The *ingi pipa* is smooth and sweet on my tongue.

Seyfu has drawn a rough map of the Marowijne River. He points to a place on the western shore. "Boni Mountain," he says. "Past Bonikreeki, where the tides stop." He is showing me where the Boni Maroons used to live, ancestors of the modern-day Aluku tribe.

Boni Mountain is haunted, says Seyfu. You must have *deki ati*, a strong heart, to travel there. Nobody has been there in more than fifty years. Now Seyfu begins to moan, softly from the back of his throat. The mountain makes noises at night, he says. If you flash a light on it, the light reflects back and burns your eyes. It is a dangerous, forbidden place.

The Indians used to help the Boni, he says. They taught the Maroons how to grow cassava, how to live in the bush. But then the Boni began stealing from the *ingi*. The Indians complained to the Dutch army in Albina, which mounted an expeditionary force. But

when the soldiers arrived at the mountain, the Boni had already fled south, to a village now called Bonidoro.

Eventually, the Boni turned on their leader and murdered him. They brought his decapitated head to Albina to show the Dutch. But as they were handing over the evidence, Boni's head sprang from their hands and disappeared into the waters of the Marowijne. The spirit of Boni still lives in this river, says Seyfu. When bad things happen, this is who the people blame.

Now Seyfu points to another sketch, a stick figure man with his hands held high in surrender. Next to this, Seyfu has drawn three arrows pointing to the man's chest. "No *feti moro*," he says. No more war. Seyfu stands up, takes another cigarette, and slowly nods his head.

A nation's history and its people's past are usually, logically, the same thing. But sometimes they can seem unrelated, one purely factual, the other intensely personal, neither able to corroborate the other. The bullet holes in the walls of the school and Seyfu's tendency to speak in metaphor are examples of this; the bullet holes in the walls of Fort Zeelandia and Bodi's reluctance to speak about Bouterse are, too.

The people of Suriname are profoundly stuck. They are trapped between the economic potential of their land and the challenge of releasing it in a sustainable, egalitarian way. They are trapped between the clamour of a worrying future and the noise of war from not so long ago. Most of all, though, the people of Suriname are trapped between a history they seem anxious to forget and a past they seem desperate to confess.

14

THE HOLE IN THE JUNGLE

The Marowijne is more than simply a political border between two nations. The river separates two continents and the Ndyuka Maroons do what they can to take advantage of this. Many men attempt to father a child on the French side of the river, so they are entitled to a monthly cheque from the French Department of Welfare, made out in all-powerful Euros. Mr. Jack tells me there is far more work for a ferryman on the Surinamese side, and the number of people seeking one-way passage is far greater in Albina than in St. Laurent. Given the militancy of French immigration policies, though, these waters are more a moat than a river.

This is not to say the French side is markedly more developed. French Guiana is, after all, merely an overseas department of France – a place famed for its historical role as a penal colony, a strategic holding rich with gold and timber, the home of Europe's largest tropical rainforest, a handy piece of real estate from which to launch rockets into space. From our boat, only small differences are apparent between the two banks. Some French houses are better built, with glass windows, concrete foundations, and shingled roofing; the roads of *les villages* are more often than not paved with asphalt; the boat launches are sturdier, cement constructions. But both sides of the Marowijne are inhabited mainly by Maroons and, therefore, resemble each other quite closely.

We pass Apatu, where the river runs shallow over wide sandbars, where jerry-rigged mining operations send the sand north to construction projects in Albina and St. Laurent. Farther upstream, steaming barges huff and puff, their pumps sucking the river bottom to the surface where labourers sift it for gold. These barges are lashed together with old towels and pieces of tarp, their decks battened down with recycled nails and ropes pulled tight. Houseboats on pontoons are similarly constructed. On the walls of one of these dwellings is scrawled the inspirational message, *"Sranan libi, Sranan lobi,"* Surinamese life, Surinamese love.

At Armina Sula, a place of roiling rapids, the clouds unleash their rains. Mr. Jack goes straight for the throat, expertly zigzagging his way through the labyrinth of rocks. Halfway up, we come across a boat lodged sideways against the current. The owner, a middle-aged Hindustani man, stands on a rock, pulling with all his might on a rope attached to the bow. In the stern sit his two children, huddled beneath a blue umbrella.

Mr. Jack beaches our boat on a shoal of smooth stones and we climb out to help. The boatman tells us his prop is gone, sheared off against a submerged rock. He is lucky the boat didn't capsize. We spread out, the water up to our chests, and grab a section of rope. We pull but the boat doesn't budge. The kids are oblivious. Both now sit astride the gasoline canisters in the hull, cowboys in a river rodeo.

Another boat climbs the rapids. It rides low in the water, weighed down with gasoline and seven teenaged Maroon boys. A thicket of wooden arms and legs, the limbs of countless tables and chairs, reach up from beneath a ratty tarp in the bow. Mr. Jack waves but they refuse to stop.

Now our boat is pushed off its shoal by the rushing water and begins sliding down the falls. Jason yells to Mr. Jack, who makes a superhero leap into the stern as it drifts past, his black rain cape billowing. He cranks the motor, flips it out of neutral, and guns it back up the rapids, crashing onto the rocks. We hear the sickening sound of wood splitting as Mr. Jack cuts the engine.

We cannot help this man and his children. If they were headed in our direction we would take them with us, but now they'll have to wait for a northbound boat. The owner thanks us for trying. Mr. Jacks says he will pray for him and his children. We make it through the rest of the rapids as water pours in through a fresh hole in the hull.

"*Watra feti mi*," yells Mr. Jack. The river is making war.

We pass through two more storms. The waters seem to rise by the second and the rapids grow increasingly turbulent. Every minute, another sharp stone drops beneath the surface to menace our propeller. Jason and I bail while Mr. Jack attacks each section of whitewater at full speed, slicing into the collapsing V of each torrent, powering through that split-second of stasis when horse-power and hydropower negate each other and the boat sits nervously still. We climb Alamukula Sula and Bonidoro Sula. We pass Langa Tibiki, or Long Island, where Mr. Jack will never return. We stop for a rest on a small sandbar, where Mr. Jack points at the red sand and mutters "Paramaribo money." Soon we overtake the boat filled with gasoline and furniture. It is creeping along, half-sunk near the shoreline.

Three hours later we reach the village of Nason, where we learn what this river can really do. The village is perched on stilts on a slim island in the middle of the Marowijne ringed by cliffs of hard-packed, pumpkin-coloured sand. During the long rainy season to come, the river will rise past these cliffs and lap beneath the buildings now towering above us.

Mr. Jack beaches the boat. We step to shore as the late afternoon sun peeks through a break in the clouds. Streams of orange rainwater pour down the cliff faces and tumble down a set of steps dug into the earth. At the top of the steps, a boy and a girl hold hands as they peer down at us.

"Can you show us to your captain?" I ask.

"*Kapten no dya*," they say.

"What about the Basia?" The children motion for us to follow.

The Basia is having a haircut.

He sits in the dark beneath a rotting house on stilts, a black cape tied around his neck. He is perhaps sixty-five, with curly grey hair and a good-sized belly. Behind the Basia stands a well-muscled man with long dreadlocks and a rainbow towel wrapped around his waist. This man holds the scissors. Neither acknowledges our presence.

Mr. Jack steps forward and launches into a formal introduction.

He begins with a long prologue, recounting the entire story of his acquaintance with Jason and me, and every detail of our journey here. He explains why we've come, that Petrus Tjappa invited us to celebrate the life of his mother; he also includes his opinion that the two white men are not here to cause trouble. The barber's chiselled torso ripples as he snips. The Basia listens intently. Every ten seconds or so Mr. Jack goes quiet and the Basia fills the silence with a rhythmic "*Ah, so.*" The discourse has a lovely lilt and poetry to it.

It takes Mr. Jack fifteen minutes to work his way up the present, in which two white men stand beneath an old house in Nason, requesting permission from the assistant headman to *koiri* in his village. Now Mr. Jack stands aside and formally introduces us. The barber stops cutting. The Basia swivels to face us. Only now do I realize he is blind, his eyes clouded over and weeping down his cheeks.

We shake hands and he welcomes us to his village. He introduces us to his barber, a man he calls simply the Rasta. Then he frowns and shakes his head.

"Thank you for coming," he says. "But the party is cancelled."

I look at Mr. Jack. He is suddenly nervous, as if he expects us to cause a scene. After all, we've come a long way and spent a lot of money to get here. Jack seems well acquainted with the wrath of disappointed tourists.

"May I ask why it's cancelled?"

Mr. Jack translates. The Basia responds curtly.

"The party will be in May."

So that's it. Now the Basia will surely ask us to leave.

"Maybe we will come back, then." I can't afford to stay in Suriname until May. We might have time to make it back to Bigiston before sundown.

Mr. Jack seconds this idea but the Basia doesn't like it. He heaves himself out of his chair.

"No," he says. "The party is cancelled, but you will stay. Tomorrow is Christmas." He points to a neighbouring house, also on stilts. "Hang your hammocks there."

"Thank you," I say, a little lost for words. We shake hands again. Then I ask one more question.

"Where can I find Petrus Tjappa?"

The Basia shakes his head again.

Petrus isn't here.

The rafters beneath the neighbour's house are low and riddled with termites and old bird nests. The wood crumbles between my fingers. The dirt floor is under an inch of rainwater and the air is thick with the sharp scent of mould. Mr. Jack returns to the Basia and asks if there is another place we might sleep. Meanwhile, an old woman approaches with an armload of mangoes.

"You walked right past me!" she yells in Sranantongo. She is very angry. "You didn't say hello! When you come to my village, you must say hello!"

"Oh, mi sari," I say. "Ma, mi no si yu."

"Yu no see me?" she says slowly, impersonating my terrible Sranantongo.

"No. Mi sari."

I smile at the woman. We stare at each other. Then she laughs.

"Teki," she says, motioning for me to take a mango from her heap. She offers one to Jason as well and then wanders off, racked with giggles.

Mr. Jack returns and takes us to the other side of the village. The houses here are all well made, with glass windows and large porches. The pathways are swept clear of debris and occasional gardens burst with colour. We approach the largest house, where Mr. Jack speaks with a well-dressed man who reclines in a rocking chair. Out front, a huge satellite dish rises above the bushes, and next door is an ancient timber-frame church. This man is the village pastor. He tells us this is the good side of the village, the Christian side, and directs us to a neighbouring house that has been empty for years.

Dusk descends as we walk back to the boat for our gear. When we return, we find a young woman mopping the floors and a young boy stringing a long extension cord through the roof. Mr. Jack puts a pot of rice on the stove and begins chopping onions. "Thank you," he says, to no one in particular. The boy flicks a switch and the house fills with light.

After dinner, explosions boom on the far side of Nason. Jason and I follow the sounds. We pass a number of doorways decorated in the traditional Paramaka fashion, geometric shapes of red, blue, yellow, and white intertwined with one another, the same designs Petrus used to paint on his immaculate carvings. We pass a house filled with Maroons transfixed by a pirated DVD playing on a flickering television set. Among the crumbling houses on the far side of the village, we find groups of children hovering in the dark, their faces intermittently lit by the sudden sparkle of fireworks. They throw them between the huts at the last possible moment and then run for cover, yelping with excitement as the blasts echo across the river.

We spend Christmas Eve on the southern tip of Nason Island. To the west, Nason Bergi rises up in the moonlight; to the south, the jungle closes in around the narrowing Marowijne. In the water, a small boulder resists the rising river. We toast our arrival with a fresh *jugo* while a swan floats past, its body a white smudge adrift in the dark. The swan stretches its neck. So does the boulder. Then both of them slip beneath the surface.

I wake to the voices of a Paramaka choir.

It is ten minutes to midnight. I creep out the front door and follow the song. Soon I'm standing in the dark outside the old wooden church, the interior lit by a hundred candles, the pews filled with villagers. They sing Christian hymns in Sranantongo. They sweep their arms from side to side while an unseen harmonica is played out of tune. The grinning preacher, dressed in robes decorated with intricate, vaguely African designs, stomps his feet and claps his hands like an old revivalist. Nailed to the wall behind him is a white porcelain crucifix, Jesus's body twisted in the familiar, snake-like pose.

The music reaches its climax. A few of the men begin to holler; a few of the women begin to cry. Then silence. The *kerki* (church) empties out. Parents light firecrackers with their children. I weave through the crowds in the dark. I might be sleeping. I might be sleepwalking.

I come face to face with the Basia. He grabs me by the shoulders and pulls me close. He stares at me as if his memory is as cloudy as his vision. He says something I don't understand, over and over. I can smell his sweet-sour breath. Then he lets me go. He stumbles away in the dark and so do his people. I spin around and realize I'm alone.

We wake to an eighties rock version of "We Wish You A Merry Christmas" blasting from a faraway speaker.

I ask Jason if he heard anything last night.

"Just some fireworks."

I ask Mr. Jack.

"*Mi go kaka heri neti.*" Mr. Jack spent the night in the latrine.

Outside, a small man limps toward the house. He goes straight to the tank of rainwater, pulls a harmonica from his shirt pocket, and rinses it off.

"*Morgu, morgu,*" he says.

"Merry Christmas," I say.

"Thank you!" yells Mr. Jack from the kitchen.

A young boy bashes a mallet against the old church bell. Soon, the villagers appear and stream toward the *kerki*. The Basia arrives. He wears a white dress shirt tucked into a pair of baby blue basketball shorts. He tells us the service is about to begin.

We find seats at the back of the church. Some of the villagers smile at us, but those we sit next to ignore us. The pastor stands behind the pulpit and opens his bible – his *Gadobuku*, the blind guide that leads him – and the congregation does the same. He speaks a few verses and then motions for everyone to stand. The people of Nason rise and the small man gives a note on his harmonica. The hymns begin. Arms are raised, bodies begin to sway, and the villagers join the pastor in song. Jason and I mouth the words. The Basia remains sitting in front of us, clapping his hands to the rhythm. The harmonica man strays wildly from the tune.

After three hymns, the music dies and a few of the villagers line up next to the pulpit. One by one, they take their turn as preacher. They pray for each other, for their village, for their country, for their earth. Each one praises the *masra* Jesus Christ, at which point the people stamp their feet and applaud. After a half-hour, the pastor returns and gives his sermon. I recognize a few of the words: *makandra* (together), *feti* (war), *alen* (rain), *liba* (river). The people listen intently. Even the children are silent. And then the pastor points to Jason and me. The villagers turn to look at us as the pastor thanks us for coming, for sharing Christmas with them. He says it makes him happy to see us here, the white man among the black man. He repeats this over and over – *weti man, blaka man, weti man, blaka man* – to insist there is no difference between us.

After the sermon, I find Mr. Jack leaving the latrine.

"Where can I find the Tjappa family?"

227

Mr. Jack points behind the church. "*Bere fukup*," he says, rubbing his belly.

"*Mi denki yu musu sribi*," I say, suggesting he go back to bed.

"*Ai.*"

"*Mi denki yu dringi tumsi biri*," I say, suggesting he lay off the beer.

"Thank you!" he says.

We spend Christmas afternoon with the Tjappas, many of whom came here for the party and were disappointed like me. We drink cashasa and fresh lime while Petrus's great-nephews and grandsons play pounding reggaeton from massive speakers. I meet Lena, Petrus's sister, whose hair is dyed bright pink. She shows us the Tjappa estate, a long field of grass bounded by an assortment of huts, dazzling *faya-lobi* plants, and key-lime trees. She tells us Petrus had to work over Christmas so the party was postponed. I show her a photo of Petrus and me and she laughs.

"You look like his son," she says.

I ask her if anyone ever visits her mother's grave; I've brought a small bottle of *palum* that I'd like to scatter there. Lena gives me a horrified look and shakes her head. I immediately realize my blunder.

"We stay in village," she says. "The dead stay in the jungle." Maroon tradition forbids visiting the grave of a loved one.

The rains come and go as we drink beneath a makeshift gazebo. The church bell rings every few hours, signalling a repeat performance. As the sun is setting, Jason and I take our leave, and Lena gives us a hug.

"I will tell Petrus you were here," she says. "He will be so happy."

After dinner, Anton, one of Petrus's great-nephews, appears at our house. We pour him a beer and Anton tells us that three years ago, his eldest brother left Nason to work in the goldbush. One day, while he was using a chainsaw to fell a tree, a piece of metal sheared off and ripped through his neck. He died in ten minutes. His co-workers returned his body to Nason two days later, their clothes stained with the dead man's blood.

"This was my brother's house," says Anton. "I helped him build it before he left. No one has lived here since he died."

We offer our condolences and thank him for letting us stay here.

"It is not me," he says. "It is the pastor. He is the one who lets you stay here."

We sit in awkward silence. Jason fiddles with his cameras while Mr. Jack paces back and forth, peeling a mango and quietly groaning. Anton fingers a matchstick thrust into the small dreads above his left ear. The rest of his head is shorn to the scalp.

Finally, Anton speaks. "You want to see the hole in the jungle?"

"*San?*"

"I can show you the hole."

"You mean the grave?"

"No."

"The hole."

"Yes."

"What hole?" asks Jason.

"*Bigi olo na busi.*" The big hole in the jungle.

"Where?"

Anton points south across the river. "*Drape. Na bergi.*" On top of Nason mountain.

He stands up and pours himself more beer. "Tomorrow I hunt there. You can come."

Early the next morning, Jason and I find Anton drinking Parbo and sharpening his machete on the other side of the village. He sits in front of a blackened stove, a round trough filled with roasting *kwak*, or powdered cassava. Flames lick out from beneath as Anton's wife shifts the yellow flour back and forth with a garden hoe. Sweat drips from her forehead and sizzles as it strikes the hot metal.

Anton runs to his hut and returns with his shotgun, an old 12-gauge, and a half-empty box of shells. Then he grabs a pocketful of *kwak* from the stove, throws the rifle over his shoulder, and leads us to his boat. We head upriver and across to the far shore, where Nason Mountain rises out of the bush. We follow Anton along an old prospecting trail carved into the rainforest by Suralco more than fifty years ago. Early on, we pass a rusted-out Caterpillar overgrown with vines and moss.

At every sound, Anton stops dead and fingers the trigger on his rifle. Jason and I hang back, unsure of how much beer he had this morning. Up ahead, a gaggle of nervous trumpeter birds skitters along the trail in single file, and far to the north a troop of bearded saki monkeys cackle. As we climb higher, the air grows cold and the forest gets younger. Huge swaths were clearcut by the bauxite hunters in the 1950s and 1960s. On one side of the trail the jungle is still so thin that I can almost see over the nearby cliff face and south to the remote Tapanahony region.

Anton stops again, and Jason and I listen. But now our guide points to the ground. A jaguar print, very fresh and the size of my palm, in the middle of the trail. "Make last night," says Anton. He's seen countless jaguars and countless pumas up here in the past.

After two hours of uphill slogging, we reach a break in the forest, a muddy two-lane road that apparently stretches all the way north to the east–west highway. We are exhausted and Anton hasn't fired a shot, so we stop to eat some *kwak*, a few Javanese peanuts, a half-rotten coil of salami. We sit at the intersection of our hiking trail and this strange road, drinking the last of the Scotch and

smoking Morellos. Tire tracks in the mud, like the jaguar print, seem to be fresh.

"How much farther?" I ask.

"We are here," says Anton. He points south down the road.

I shoulder my pack, leave Jason and Anton behind. A few hundred metres away, I find the hole. It is a rectangular prospecting pit, roughly ten metres wide and four metres across, surrounded by rice sacks filled with lucrative red earth. The remnants of mighty lianas dangle from the canopy above, their lower portions sliced off. The understory has been hacked, piled, and burned nearby. There is bauxite beneath this forest, perhaps across the entire Nason plateau. I am staring into the grave of a jungle.

I reach into my pack, retrieve the small bottle of *palum* I'd brought for Petrus's mother. I unscrew the cap, take a short swig, and scatter the rest into the hole, just as Petrus had taught me at Raleighvallen. I mutter a few words of blessing and then return to the others. Both Anton and Jason are a little annoyed to hear I've wasted the rest of the booze.

We descend Nason Mountain the way we came. Anton manages to get one shell off, an ambitious, futile shot at a *powisi*, a black currasow, perched high in a thicket of vines. Back at the river, as we're pulling away from shore, I spot three white men on the French side dressed identically in tight camouflage tank tops, dark green short-shorts and black ankle boots. They are frolicking on a short wooden dock, pushing each other toward the water. One of the men leaps up onto the back of another man who begins to spin in circles. The rider's legs splay out and I think I hear him scream. The pale flesh of his inner thighs shimmers in the sun.

I ask Anton.

"French soldiers," he says.

I look at Jason, whose mouth opens in a devious smile. Without a word, Anton swerves across the water to France.

Lieutenant Gavalda of the 13th Regiment of Army Engineers, French Gendarmerie, greets us at the gates of the base.

"Come in, come in," he says happily, shaking our hands. "We enjoy visitors. Visitors are nice."

Gavalda is the boss here but you wouldn't know it by looking at him. He is only twenty-five and wears the same skin-tight uniform as his soldiers. He has a broad physique, bright, expectant eyes, and legs that are so thick his shorts must be cutting off circulation.

"*Alors*," says Gavalda. "Welcome to paradise."

We stand on a wooden boardwalk just inside the gates. The walkway weaves through the base, which resembles a traditional Native American fort. Thatched-roof buildings are separated by lush gardens and manicured lawns. The whole camp is enclosed by eight-foot walls of upright timbers. Soldiers busy themselves tending gardens or hauling water or scrubbing floors. Lit cigarettes poke from their mouths and packs of smokes bulge from their breast pockets.

"The boys are desperate to swim," says Gavalda. "We are from Besançon. We are not accustomed to this heat. Sometimes I worry they will jump in."

"Why worry?"

"Because then I will have to discipline them."

"For swimming?"

"Against regulations. Gendarmerie is not allowed to swim in rivers. Too many risks. Sharp rocks, currents, piranha, electric eels." Gavalda frowns. "I do not like to discipline. It makes the boys unhappy."

"So how do you cool off?" I ask.

"Ah," says the Lieutenant. "We have next best thing. Our bath-houses are the best on the Maroni. We can shower with a view of the rapids. Sometimes the men stay in there for an hour or more."

The 13th has been here for three weeks. They will stay for another three months. Then they will be replaced by another detachment of French cadets fresh from the *académies service*. Their goal here is simple.

"We count Brazilians. We do not arrest. Just count."

Every week, a group of soldiers travels upriver to the Marowijne goldbush. They visit the camps and count the number of *garimpeiros*. Then they pass this information on to officials in Cayenne.

As we talk, Gavalda's boys wander past on the boardwalk, their boots clacking. They all seem very friendly, more like cabana boys than soldiers.

"Can we see the rest of the base?" I ask.

"Ah, *désolé*. Against regulations. You must get papers in St. Laurent. But believe me, it is paradise. We sleep very well. We eat very well. Our chef, he is from Martinique."

Jason, silent up until now, decides to speak.

"Brothels," he says.

Gavalda squints at him. "*Quoi?*"

"Whorehouses," says Jason.

Gavalda still doesn't understand. Neither do I. I give Jason a look he later describes as "What the fuck?"

"Sorry," says Jason, fiddling with the camera around his neck. "I don't normally ask the questions."

"I think my colleague here is asking about the prostitutes in the goldbush."

"Ah, *les bordels!*" What do you want to know?

Jason think for a moment. "Well," he says. "Do you count them, too?"

"No."

"Why not?"

Gavalda grimaces. "Let me say, they are not our operational intent."

I give Jason another look, which he later describes as, "Satisfied?" He responds with one of his own, which I interpret as, "It's called asking the tough questions. It's called journalism."

We thank Lieutenant Gavalda and he leads us back through the gates. I stop in front of a tall wooden post cluttered with colourful road signs, one for each French regiment that has toured here: the 8 RA from Commercy, 7,459 kilometres away; the 35 RAP from Tarbes, 6,823 kilometres away; the 601(2) RCR from Arras, 7,601 kilometres away; two regiments from Hussards-Sourdon, 7,536 kilometres away. All the signs point northeast.

In the morning, Mr. Jack takes us to say goodbye to the Basia. We find him sitting on a dilapidated porch, cradling his five-month-old grandson. Mr. Jack immediately launches into a long account of our visit here while the Basia inserts the occasional "*Ah, so.*" Mr. Jack thanks the Basia and so do we, handing him a bag of leftover groceries. Then the Basia passes the baby to his wife and addresses us.

"My side of the village," he says, pointing to the decrepit housing around him, "is very poor. Not like Christian side." Only now do I realize his authority does not extend to the other side of Nason. "You come to Nason, you live with the rich. If you come back, you live with us."

The trip north is fast and beautiful. The sun is out, the river is full, the rapids have smoothed out; we shoot them without thinking, Mr. Jack's attention to detail long gone. We stop three times on small sandbars so he can leap out and crouch between the shrubs.

On the third stop, he returns with an armload of young *guyaba* leaves. He will boil these when he gets home and bathe in their healing extract.

We stop in the French town of St. Laurent before returning to Albina. Mr. Jack waits by the boat as we walk along Avenue Félix Éboué, where Peugeots and Renaults fill the street, where well-tanned Frenchmen in button-down cotton shirts smoke Gauloises Blondes in rundown cafés. We change our remaining money into Euros and stop at Restaurante Chez Titi. Tables of European families and gendarmerie spill out onto the patio. We order a takeout lunch in a ridiculous mixture of French, English, Spanish, and Sranantongo while our waitress, a young Ndyuka who is fluent in five languages, lists off the Belgian beers and Italian wines. We order two Leffe and a *pizza à trois fromages*, Roquefort, Camembert, and Emmenthal. Then we hurry back to the boat.

We offer Mr. Jack some pizza but he frowns and cranks the motor. In ten minutes we are back in Suriname. At the Albina Breeze, the air is filled with moaning. "Romance is a dirty word," says Jason, as we listen to the prostitutes hard at work.

The next morning, Mr. Jack arrives for his tip. We are out of money so we give him an LED lantern. Jason tries to explain the miracles of LED but Mr. Jack frowns. He'd wanted money; he has no use for gadgets. We promise to send a tip back with his son, who will drive us back to the city. "Thank you," says Mr. Jack, finally getting it right.

We average 160 kilometres an hour on the east–west highway and are soon cruising across the nameless bridge into Paramaribo. The city looks quiet again, charming, free from its history and its past. This is the bridge from which a number of despairing townspeople have leapt to their deaths this year.

Romance *is* a dirty word. Like all travellers I am anxious to avoid it, to push past softness and cliché and discover something fundamental, hard, and true. But now sadness has crept into everything around me. I can't help thinking of Voltaire's *Candide*, the man who lost all hope here, the man for whom Suriname "plunged him into a black melancholy" in which "he fed only on sad ideas."

I call Emma from the pay phone at Stadszending. Her answering machine picks up. Over an indie rock soundtrack, she uses her middle name. She never uses her middle name.

15

―

THE RED CITY

This morning the walls are shaking. My Scotch quivers in its glass as the bombs land in the streets. *De Ware Tijd* describes the latest models, imported Chinese firecrackers shaped like robots and AK-47s, toy ordnance aimed squarely at Suriname's children. Every year at this time, the young lose fingers and eyes as Paramaribo is engulfed by a warlike party.

"Domineestraat is on fire!" yells Jason, bursting into my room. For the next few days, travel into the interior will be impossible – today is the last day of the year, and the celebration is in full swing.

Domineestraat is packed and lined with beer tents. A reggae band howls from a makeshift stage as the crowd sways, arms raised in triumphant fists or waving open *jugos* and bottles of Red Label. Latin and Caribbean dancers grind on the stage roof. Drum squads march through the intersections and pound their massive skins. Creoles, Hindustanis, Javanese, Chinese, and everyone in between wrap arms around each other and drink themselves into oblivion or the New Year, whichever comes first. The entire city is out on the streets, two hundred thousand people, and the balconies of the Krasnapolsky Hotel are jammed with tourists, Dutch and German, watching the locals party.

The rains arrive. We buy two *jugos* and sneak up the back stairs of an Old Dutch colonial building. On the terrace a well-heeled family is observing the mayhem with nervous smiles. The people below scatter. Children scream and adults cover their ears. A strange crackling rises on the air, then an explosion rips through the street.

The New Year's celebration in Paramaribo is possibly the world's wildest display of pyrotechnics. The government spends millions of dollars each year importing the explosives. Rumour has it every December 31 more fireworks are detonated here per capita than in any other city on earth.

Acrid red smoke engulfs the balcony as the music returns and the crowd below roars. A storefront across the street vanishes in a cloud of red as a long string of M80s cracks off. Then another, and another, all the way down Domineestraat, the fireworks booming and sparkling like tinsel. There is no rhyme or reason to the explosions. The only way to know where the next might occur is to watch for a sudden break in the crowd with a brave man kneeling at its centre, his shirt held over his face, his hand out, his thumb flicking a lighter. Then he, too, runs for cover, and another string of bombs goes off. These are the original M80s, first designed by the U.S. military to simulate the guns of war. They have been illegal in the U.S. for decades and are the beating heart of the Surinamese New Year.

"Here's to postmodernism," says Jason, clinking my *jugo* with his. We return to the street and wade into the throng.

For the next four hours, we stumble and stagger through the rain, drinking hard and trampling each other in the red fog, screaming with fear and adrenaline as each new chain of shells is set off. At times the smoke is so thick I can barely breathe. Between blasts, terrified children wail, the drums bang, the reggae pounds, the booze flows. Every half-hour or so, a chain of firecrackers terminates in a massive explosion, a globe of bright red fire showering the street with flames. Domineestraat is a cesspool of mud and bottles and red tickertape, shrapnel from the M80s staining our shoes and turning the gutters crimson.

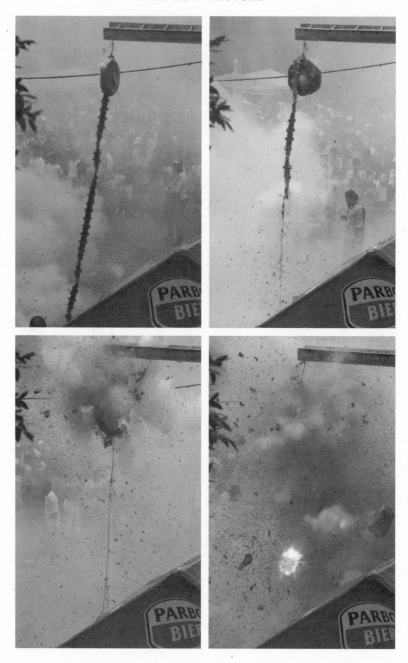

Now an ominous box-truck inches down Domineestraat. From its rear doors, volunteers uncoil a half-mile-long chain of ordnance along the street. The process is excruciatingly slow and works the crowd into a frenzy. Finally, a young man steps to the lead end of the chain, lets loose a triumphant howl and touches his lighter to the fuse. The smog glows red as the screaming fire creeps up the street.

We bounce from party to party all afternoon. In the tourist district, we sit at the T'vat bar with a few Peace Corps volunteers. I meet Jill, from Missouri, who lives in Drepada, just upriver from Balingsoela.

She's soft-spoken with bright blue eyes and tells me about good things, the hallucinogenic *ayahuasca* she bought online and burned in her kitchen, the meditation she uses to calm her mind. She sees the tattoos on my forearms but doesn't ask. She tells me she's headed to Mali next week to meet up with a friend.

Night falls and the party moves to XPO, the only gay bar in Suriname, where Jesse, Fabian, and Mario, the Number One DJ in

Suriname are spinning. Jill and I dance tightly together. We are a fine fit. My hands find her body slowly, secretly, the only straight hands in the house. I get drunker and so does she. Then she pulls away in the dark. She says one word, *trouble*, and I know what's coming next. Jesse is spinning. He's improved in just one month. I keep drinking. Jill and I dance again and she says it again, *trouble*, and I ask her what's wrong. *I'm trying to be good*, she says, and I imagine her man in Mali trying to be good, waiting for her plane to arrive. I become more physical. My hands and feet lose their grace as everyone starts to bump and grind. I begin to embarrass myself. I do it slowly, the worst way to embarrass yourself. I grab her hand three times and three times she says no by letting go. Mali is in northwest Africa, just below Morocco. Toronto is in southern Canada, on the shores of Lake Ontario.

I want to be honest with her, to ask her to fuck, to tell her what's happening to us is natural and normal. Because that's what Emma told me on the phone – *everything that's happening to us is natural and normal*. I quote her as I try to get close to another woman.

I reach that breaking point. I lean in to speak in Jill's ear. *She broke up with me two weeks ago.* Jill looks up. She might be about to wrap her arms around me. She might be about to soften. Everything inside leaps. Then she shakes her head and walks away.

An hour later, Jason and I are slumped in a taxi, racing through clouds of red, yellow, and blue, dodging the blazing missiles that strafe the neighbourhood streets. We're late for a party at the home of Jack Fernandes.

Inside the Fernandes front gates, on an outdoor patio and gathered around a gourmet buffet, every member of the Peace Corps is on their best behaviour. Two years ago, this party went out of control. Last year, the Corps were not invited back. Now, Fernandes has

softened and lifted the ban, but everyone is well aware of the diplo-
matic stakes. Jack Fernandes is a very powerful man, one of the
richest people in Suriname. The Fernandes Group sells everything
from Japanese cars and electronics to pastries and ice cream and
is Coca-Cola's Surinamese distributor. When John F. Kennedy
inaugurated the Peace Corps on March 1, 1961, one of his goals was
to combat the growing anti-American sentiment throughout the
world, characterized by the idea of the Ugly American. Perhaps
chastened by their comrades' debaucheries two years ago, a few of
the Peace Corps guys are actually wearing ties.

Unfortunately, the unmanned liquor table is help yourself and
impressively well stocked. It's being eyed by forty-odd jungle-
dwelling Americans and two uninvited Canadians. We've all been
drunk for ten hours. Pretty soon the ties are off and a boisterous
posse has gathered around the bottles.

I sit and drink for a while with a middle-aged woman and her
puppy-dog husband, who laughs at everything we say. Halfway
through the conversation I realize this woman is the American
ambassador. I ask her about the controversial plans to dig up Dr.
Sophie Redmondstraat to expand the American embassy, plans that
involve uprooting a much-loved statue dedicated to Suriname's
children and moving it across the street.

"It's simple," she says. "Security in the twenty-first century means
beefing up our embassies." The ambassador smiles. Her husband
barks at the mention of meat.

At midnight, we drink champagne poured over a pyramid of
glasses on a patio shrouded in sour smoke. Music blares from hidden
speakers and the volunteers begin to dance, months of pent-up
sexual energy adding to the fog. Jason finds me sitting on the edge
of the pool in the backyard. I have no idea how I got here. He holds
a giant cup brimming with Coke and Black Label.

"What are you doing?" he asks.

"I can't remember."

Jason hands me the cup. "Let's go down the rabbit hole."

I follow him into a small coach house. Inside is an immaculate children's playroom filled with hundreds of plush toys, two rocking horses, a wrought iron bed, and a costume trunk. On a tiny table surrounded by tiny chairs is a miniature tea set for eight. Two Peace Corps girls are sitting on the bed, two giant teddy bear heads at their feet.

"It's like Alice in Wonderland," they say, handing us the massive heads. The night takes a strange turn.

The view from inside the heads is through a piece of see-through fabric over the bear's nose. Through this lens the playroom becomes surreal, detached, as if I'm watching a stilted home movie of my childhood had I been born a girl in 1950s America. It is vaguely frightening and absolutely hilarious, and soon I'm laughing hysterically. I hear Jason laughing, too. When I turn to face him a giant chimera – half human, half bear – stumbles past my nose-screen.

The next hour passes like a hallucination. We slam into each other, bounce off the walls, wrestle each other to the ground. My laughter grows manic, and my stomach begins to burn. Something breaks and everything starts to shine. For one endless moment, I think about absolutely nothing. We crash into the table, sending the tea set flying. We ride the rocking horses, sending each other into convulsions. Somehow we end up on the bed, the mattress thrown clear, the iron frame creaking, collapsing, our asses stuck through the springs. We remove the heads, tears streaming down our faces, two sweat-soaked Ugly Canadians. The girls are long gone and the party is winding down.

The New Year in Suriname has officially begun.

16

—

THE NEW POLITIC

Two days later, Jason's last in Suriname, the streets of Paramaribo are still littered with muddy piles of red tickertape. I meet Bruce Hoffman bright and early at the Zeelandia Coffee House.

"This 'pristine forest' thing is bullshit," he says, as an exhausted waitress delivers our Americanos. "Ecologists worship the word. So do tourists – no one wants to spend thousands of dollars to visit a 'nearly untouched' jungle. But they need a different word. The forest is constantly changing. The forest has never been 'pristine.'"

Bruce has spent many months over the last few years in the jungles of Suriname. He will soon complete his Ph.D. in ethnobotany, based on his fieldwork with the Saramaka Maroons at Awaradan and the Trio Indians near the border with Brazil.

"And the forest isn't 'empty,' either," he continues. "Some groups say it's empty, to make conservation policy less complicated. But one look at a Maroon resource map and you realize that's bullshit, too. Fishing spots, hunting spots, burial sites, *maripa* nut gathering sites – everything in the forest has value and meaning."

Bruce works with the Amazon Conservation Team, the non-profit cultural preservation organization founded by Mark Plotkin after he left CI in the mid-nineties. Hoffman could be Plotkin's younger self, a scholar-adventurer obsessed with documenting the medicinal properties of rainforest plants before the Maroon medicine men and Amazonian shamans disappear forever.

Bruce glances at our Americanos: two shots of espresso and eight glasses of steaming water. "I love this place," he laughs.

"I met Fritz," I say.

"Oh yeah?"

"He's still pissed."

"At Mark?"

"Yeah."

"Figures."

Bruce tells me the real reason Fritz von Troon was fired. Mark had hired him to oversee the construction of a new eco-lodge, paid for by the World Bank. Two weeks before completion, Fritz disappeared.

"Then one day we saw him on the cover of *De Ware Tijd*," says Bruce. "He was in The Hague, accepting the Order of the Golden Ark."

"Fritz said he'd help me get to Kwamalasamutu. He said he'd radio the Granman for me."

Bruce smiles. "Good luck with that."

"I'm calling him today."

I search my bag for the scrap of paper with Fritz's number on it. I grab my cell phone and Bruce smiles again. "Go for it."

It rings five times. Suzanne von Troon picks up.

"Andrew!" she yells.

"Is your father there?"

"Fritz *gwe na busi.*"

"When's he coming back?"

"Maybe April."

"When did he leave?"

"Three weeks ago."

I hang up.

"Fuck."

Bruce laughs. "Why do you want to go to Kwamala?"

"I want to find *okopipi*."

"Why?"

"I just do!"

"Did you ask the minister?"

"Yeah."

"No permit?"

"Fuck!"

I'm out of options. Fritz was my last chance. I'm not going to see *okopipi*.

"You need a permit," says Bruce.

I glare at him.

Jason and I toast his departure with a late brunch at the Torarica Hotel. We sit in the sun-drenched dining room in our mud-caked clothing and devour the omelettes, frittatas, waffles, sausages, and guava juice of the rich. We are surrounded by elegant women and millionaire men, Caribbean celebrities from film and music, foreign families with Roman Catholic quantities of children. Neil Diamond's *The Christmas Album* blares from a hidden speaker.

I complain to Jason about *okopipi*, how disappointed I am that I can't go see it. This tiny blue frog was just about the only thing keeping me in this country. The irony is almost unbearable: thanks to this foolish *koiri*, I don't have much reason to go back to Canada, either.

I look up. Jason is staring at me through the hole in a peeled mandarin orange.

"So this is what it feels like to be sober," he says.

I drop off Jason at the Zanderij airport. Then I meet Stadszending's newest tenant.

Her name is Monique. She is nineteen years old and very beautiful, with long black hair in ringlets and perfectly round cheekbones. She has just arrived from Moengo, speaks excellent English, and is an elected member of Suriname's Youth Parliament. She is in town for a meeting at the Palace. I sit outside her room and chat through her open door.

"The solution to Suriname's problems is easy," she says, as she irons an immaculate, cream-coloured ball gown on her bed. "The National Assembly is filled with old men. Parliament should have an age limit."

"*Oude rotten*," I say.

"Exactly."

"You think teenagers should be allowed to run?" I ask.

"Absolutely!"

I laugh. "And what do you want to be in parliament?"

"A minister. Then president."

"Oh yeah?"

"Yes! Suriname needs a president from Moengo. Someone who will stand up for the Maroons."

She says nothing about being a woman. I like this girl.

"You shouldn't smoke," she says.

"I know."

"You shouldn't drink coffee, either."

"Yeah . . ."

"Or tea."

"Come on . . ."

"And tattoos are stupid."

"Hey!"

Monique laughs. "They are! They mean you don't like changing your mind!"

I just watch her iron, try to enjoy my Morello.

"What do you think of Brunswijk?" I ask.

"He's a good man."

"Because he's from Moengo?"

"No. Because he's my uncle."

Monique has the Jungle Commando in her blood. No wonder she has opinions.

"OK, advice," she says, holding three belts up to her dress. "Which one?"

"The red one."

She scowls. "That's the worst one!"

"Then why'd you hold it up?"

"Because," she says. "It's like politics."

A motorbike howls into the courtyard.

"I need favour," says Bodi, lifting his bike from the dirt.

We weave through the back alleys of Paramaribo, passing rotten slave quarters still standing after centuries of disuse, historical land-marks that in the West would have been restored and opened as haunting museums. We emerge onto Rode Kruislaan and motor onto the grass of a darkened neighbourhood park. Bodi stops beside a rickety see-saw and pulls two Parbos from his jacket pocket. He cracks them open with his teeth as he surveys the area to make sure we're alone.

"I need something."

"OK."

"The numbers one to twenty-six. How many different ways can you write the numbers one to twenty-six?"

I don't understand.

"You have paper?" I hand him my notebook. Bodi writes numbers across the top of a page. Then he writes the same numbers in a different order. "How many?"

"I have no idea."

249

Bodi smirks as if he doesn't believe me. It's the same look he gave me months ago during a pickup soccer game, after he slipped the ball between my legs.

"I'm not a mathematician, Bodi."

"Can you find out with computer?"

"Maybe. Why?"

"My business, *toch*."

"Your business."

Bodi frowns. "I will make much money. We both will."

"How much?"

"I make forty thousand. You make ten. America dollars."

"Who will pay?"

"Doesn't matter."

"Yeah, it does."

"Big Man pay, *toch*. Can you do it?"

"I don't know."

"I will tell you what you want to know."

Bodi's skin glistens in the light from a nearby Fernandes supermarket sign.

"I give you combinations, you tell me about Bouterse?"

"*Ai, toch.*"

"How do I get the money out of the country?"

"Your problem."

"Pretty big problem."

"Not too big." Bodi smiles. "As long as you're quiet."

"And what if I'm not quiet?"

Bodi shrugs. "They kill you."

Monique is still up when I get back. Her dress hangs on her closet door, the red belt wrapped around its waist. I sit on one of her beds.

"My parents don't want me to have a boyfriend," she says.

"Probably wise. You're young."

"I believe in *Gado*. Do you?"

"No."

"*Yu no go kerki?*"

"No."

"What do you believe in then?"

"Something." I wave my hands in the air.

Monique grabs my hand and rubs it. "*Gado* loves you. *Gado* cares for you. *Gado* gave you life and *Gado* can take it away. You must believe in him."

I shrug.

"Do you believe in Jesus Christ?"

"No."

"I believe in the Father, the Son, and the Holy Ghost. You should, too."

"The Hindus don't."

She hits me with her pillow. "They should!"

The air conditioner hums to life. Monique hangs her dress in her closet. Then she flips the light off and jumps on top of me.

"*San yu wani?*" That question again.

"What?"

She kisses me. "*San yu wani?*" Then she pushes the pillow over my face and her body into mine. She grinds her hips into my stomach. I can barely breathe. I catch glimpses of her face when the pillow slips; her eyes are closed in concentration. All she understands about sex is there must be a clashing of bodies.

"Monique?"

She lifts the pillow, kisses me again.

"This is not a good idea," I say.

"Why?"

"You're young. You don't know me."

Monique stops moving. "I've never had sex."

"That's OK."

We lie in silence for a time.

"I'm going to cry," she says.

"You have nothing to cry about."

She starts crying. "You can't tell anyone. Please. My mother would kill me."

"And your uncle would kill me."

She laughs. "Promise."

"I promise!" And then I tell her she is like a deer.

"*San?*"

And for the first time I understand culture. Monique has never seen the movie *Stand By Me*. She doesn't know about the deer Gordie saw one morning on the tracks, the deer he told no one about, his only secret as he and his friends searched for the bones of Ray Brower. He knew that deer was important but he didn't know why.

"It means I promise not to tell."

Monique slips off the bed, flips on the light. "Can I borrow your mobile?"

"Who are you calling?"

"My mother. My brother."

I go outside for a smoke, two smokes. I wonder if Ronnie Brunswijk is an understanding man; I wonder if he remembers what it was like to be young. Monique invites me back inside. She lies down on one bed and I lie down on the other. I wait for her breath to even out. Then I get up, turn the lights out, and leave.

Young Monique from Suriname. Perhaps everything I'm looking for is in this one woman.

The next morning, I email a mathematician friend, then take a bus to Nickerie, on the eastern border with Guyana. I wander the swamps of Bigipan, Suriname's largest lagoon and the world's largest nesting ground for the scarlet ibis. My boatman, a skinny Javanese hunter named Minh, guides us through the riddle of dead trees, a surreal avian sanctuary where the snags and branches are spotted with hundreds of brown pelicans, brown boobies, tropical cormorants, boat-billed herons, Maguari storks, Caribbean flamingos, and the occasional magnificent frigatebird. Then Minh asks if I want to see the ocean. He beaches the boat and we walk to the coast, a two-kilometre slog through thigh-high mud. I lose my sandals to the bog. I open my bare feet on the spines and buried bones of long-dead trees. Thousands of miniature snorkels – the mangrove's innovative roots – reach up from the grey mud, gasping for oxygen.

Minh stops, tells me to be quiet, claps his hands above his head. And suddenly, all around us, a glorious red cloud erupts from the surface of the lagoon. There are more than ten thousand nesting pairs of scarlet ibises in Bigipan. The hundred or so above us wail, wheel, and circle, then make for the open ocean. Minh and I follow them.

We are on the lip of the Wild Coast. From the edge of this mangrove jungle, a long tidal mudflat stretches to the horizon. In June 2000, a Canadian oil company, CGX Energy, nearly started a war between Suriname and Guyana while exploring a triangle of contested waters just north of here. This maritime border region contains the second-largest untapped oil reserve on the planet and has been in dispute since colonial times. Less than thirty-six hours after CGX had set up their rig (with Guyana's blessing), Surinamese gunboats laid siege to the operation, evicting the crew and towing the rig away. The political furor that followed brought the two nations to the brink of war. The case went all the way to the United Nations International Tribunal for the Law of the Sea, where Guyana, and CGX, recently prevailed.

I imagine these tidal mudflats blackened by an immense oil spill, the Bigipan swamp contaminated with millions of barrels of raw crude, thousands of rare and remarkable birds dead or poisoned or pushed to the brink of extinction. Then I ask Minh what he thinks of the scarlet ibis, the majestic bird that hunters are forbidden to shoot.

"Tasty good," he says, further breaking my heart. "Tasty, tasty good."

Back in Paramaribo, I find a large spreadsheet of numbers in my email inbox. I burn it to CD and delete the message. The next night, Bodi appears at Stadszending with a shovel strapped to the back of his bike.

My tiny room feels suddenly smaller as I hand him the CD.

"If Big Man wants, he pay," he says. "I go Albina for duty. Then I call you."

"How long?"

"Three weeks."

"Why so long?"

"Trust."

"I might be in the bush."

"Where?"

"The Upper Suriname. I'm leaving tomorrow."

"I wait for you. I call you. You must trust, *toch*."

I walk Bodi down to his bike.

"What's the shovel for?"

"Big fire two days ago. I dig for bottles there. Many old bottles under Paramaribo buildings. White like bones. Dutch bring them over with slaves, filled with gin. Tourists pay *furu moni* for them."

"What do the tourists do with them?"

Bodi shrugs. "Take them back to Holland where they started."

17

—

THE SARAMAKA NICODEMUS

At the border town of Atchoni, I wait out the rains beneath an
aluminum shelter. On the far side of the road, the red star of
an abandoned Texaco station looms through the murk. I can just
make out the rank of *wagis* in the middle distance, dripping bauxite,
but the boats moored at the dock have vanished. It is mid-January,
a near-sighted time in the tropics; the early, short rainy season has
wrapped this country in an aqueous veil.

A Saramaka boy of seven sits next to me. He holds a plastic beer
cup to the curtain of water streaming from the roof, and when the
cup is full he passes it to me. I pour the rain into my flask and pass
the cup back for more. When my flask overflows, the boy laughs
and pours the next cup over his head. "A *kowru!*" he yells. It's cold.

An hour later, I am almost asleep when an old Dutchman sits
down across from me. His face is pale, his body pear-shaped and
fleshy, but his voice booms through the storm.

"Have you found zee Lord Jesus?" he howls.

"Uh, no," I say. The young boy at my side is gone.

The man smiles. "Zen what are you waiting for?"

"I'm waiting for a boat."

He frowns. "It is not nice to mock."

I stare at the man.

"We are all waiting for our boat," he says.

"Yes. No. I'm headed upriver."

The man stares at me for a moment. Then he heaves himself up from the bench and claps me on the back.

"Zee Lord will be your boat today. My name is Wout. You can ride with me."

I have come here to meet the king.

Atchoni is just south of Brokopondo Lake, where the waters of the Suriname River empty into the reservoir. Cut off from the coast by at least a day's journey and bounded by impenetrable jungle, this area is the gateway to the remote Upper Suriname region, where the majority of Saramaka Maroons live. For the next hundred and twenty kilometres the riverbanks are crowded by more than sixty villages, many founded by the first Maroons to escape the plantations two hundred years ago; other villages were built less than fifty years ago by refugees who fled south during the flood.

This is the land of *Fesi-ten*, First-Time – sacred tales of rebellion and escape that have been passed for generations in the Saramaka consciousness and preserved by Richard Price in his 1983 book, *First-Time: The Historical Vision of an African American People*. These creation stories are still used to establish Saramaka land rights and to settle political disputes between clans. The stories also prove, as Price writes, that "Saramaka collective identity is predicated on a single opposition: freedom versus slavery."

Deep in the jungle, at the headwaters of the Suriname, lies the village of Asidonhopo. It is here, where the Pikilio and Gaanlio converge to form the great river, that the paramount chief of Saramaka, Gaama Belfon Aboikoni, makes his royal home. After spending so long travelling in his peoples' lands, I want to visit this village, known as Gaamakondre, to ask Aboikoni a simple question.

Wout is a Pentecostal pastor from the Netherlands. He has been preaching in Saramaka, on and off, for more than thirty years. Nearing seventy, Wout has returned to the Upper Suriname to visit the numerous villages his mission helped convert. He is accompanied by his friend Willem, who works for an evangelical television network in the Netherlands, and an energetic young Christian couple, Sheila and Walter, religious activists from Paramaribo.

In an hour, the storm lets up and we pile into the boat. Nearby, a scrawny Rastafarian picks through the trash lodged in an old fishing net. Denny, our boatman, yells at him.

"What are you doing?"

The Rasta looks up and waves. "I clean the jungle!"

Wout leans forward and taps me on the shoulder.

"You see zis man?" he asks, pointing at the Rasta. "Zis man is like you. He is searching for something but he doesn't know what it is." I turn and flash Wout a smile but he's not smiling. "Zis man is waiting, waiting, waiting. He thinks he is living but he is not. He has not even been born. Sometimes, brother, you have to make a choice."

I tell Wout about the man I saw this morning, a dying man on the road to Atchoni. He was lying on a blue tarp next to his *wagi*, which was parked sideways in the middle of the road a few hours south of Brownsweg. His fellow passengers had gathered around him, and as we approached they waved us down. The man is barely breathing, they said.

There was nothing we could do and everyone knew it, but before pulling away, our driver reached between his legs and retrieved a bottle of Fanta orange. He handed it through the window to one of the onlookers, who took it to the old man, gently tilting his head back so he could take a sip.

Wout listens to my story with a blank face. When I'm finished he smiles.

257

"A taste of sweetness before death," he says. "Zis is why I am here."

The southern reaches of Saramaka are where the rebel slaves and the Dutch finally made peace in 1762, after almost a century of war. But that's the conclusion to the story of First-Time. The beginning took place in 1690, in Jews' Savannah, north of here, when an escaped slave named Ayako led a rebellion at a plantation on the Cassewinica Creek.

This rebellion was the first slave revolt in Dutch Guiana, an event Richard Price calls the Great Raid. Ayako launched the attack from a hideout on the Matjau Creek, a minor tributary of the Suriname River (the word "Matjau" is derived from the name of the man who owned the doomed plantation, Imanuel Machado, a Portuguese Jew). Ayako and his fellow rebels set in motion nearly a century of bloody revolts and silent escapes throughout the colony's interior. These courageous slaves have been immortalized in First-Time as founding members of the fierce Matjau clan, the most powerful clan in modern-day Saramaka, the clan comprising the matrilineal line of the Gaama. All Matjau claim to be descended from Ayako and his compatriots, the "first to walk upstream."

Price relates the story as told to him by a Matjau named Otjútju: "So they attacked. It was at night. They killed the head of the plantation, a white man. They took all the things, everything they needed. And then they sacked the plantation, burned the houses, and ran."

The battle for freedom had begun.

The rain resumes as we motor upriver. The missionaries huddle beneath a tarp but there is no room for me, so I sit in the bow and shiver. Through the downpour I glimpse occasional groups of Saramaka women on the riverbank, slapping clothes and scouring dishes. Behind them, thatched roofs hide among the trees.

When we reach the village of Abenaston, our destination for the night, Denny beaches the boat in front of a massive mango tree. The shore is deserted but for a small boy fishing among the rocks. Behind him, an ancient timber-frame church stands resolutely in the rain, its walls sagging at the corners, its steeple red with rust.

Abenaston is a Christian Moravian village. It was built long before the Afobaka dam was closed and was one of many villages to welcome refugees fleeing the rising waters. This mass migration south mirrors the migration, 250 years earlier, of the countless slaves who escaped their masters. It is also the direction that the logging companies are headed. Their concessions incorporate almost every village in Saramaka. One of these companies plans to build a road just behind Abenaston that will unlock the southern jungles of Saramaka to foreign loggers, many of them Chinese – every day, more of the Saramaka heartland is felled, shipped to Paramaribo, and exported.

The boy helps us carry our gear up to a nearby guesthouse. Then he agrees to take me into the village. As I'm leaving, Wout offers me a room in the house.

"You must stay with us. You will be . . . how should I say . . . more comfortable here."

"I have friends," I say. "I have a hammock."

Wout looks at me as if I'm his wayward son.

"It is OK," he says finally, breaking into a broad grin. "You have nothing to fear here. But you must come to my service. Six o'clock."

"I'll see you then."

I follow the boy up a short hill and across an overgrown soccer field. The rain has stopped and the villagers are beginning to emerge from their houses. I nod to a few of them and they nod back. I'd lied to Wout. I don't have friends here but am hoping to make some soon.

We reach a small cluster of huts and the boy points to the nearest one. Beneath the thatch overhang, a lone figure swings in a hammock.

"*Badja-mai!*" the boy yells.

The hammock goes still. Arms and legs appear from the folds of cotton, followed by the startled face of a white girl. She looks very confused for a moment but quickly figures me out.

"You're the Canadian," she says. Apparently, word of my travels in the jungle has spread throughout the Peace Corps.

"Yes," I say.

"*Abena-folo!*" she yells.

"What?" calls a voice from the neighbouring hut.

"It's the Canadian!"

Rachel and Ariel, or *Badja-mai* and *Abena-folo*, have lived in Abenaston for a year and a half. Their development project – a pump system that provides fresh drinking water to the whole village through conveniently placed faucets – is almost complete. Both are very happy here but are looking forward to going home.

"In a way, we'd never have been able to come here if it wasn't for King," says Rachel. "Wait – how do you say 'dream'?"

Today is Martin Luther King Day in the United States. Rachel is translating the "I Have a Dream" speech into Saramaccan. The girls plan to present his speech to the children later tonight at church.

We sit around Ariel's kitchen table. Her house is larger than other Peace Corps houses I've seen. The interior is decorated with colour photographs, full bookshelves, and artwork by local children. There is a warmth to the place, as if Ariel has lived her whole life here.

I produce a package of cookies. Rachel's eyes grow wide but Ariel politely declines. She has celiac disease. Wheat gluten gives her horrible stomach aches and causes her skin to blister. It can be deadly if left unchecked, especially when the victim is living deep in the jungle and far from help.

"What are you going to miss most when you leave?" I ask.

"The people," says Rachel. "But not the men."

"Why not the men?"

"The double standards," says Rachel. "Men can do what they want but women can't. Hey, how do you say 'self-evident,' as in, 'we hold these truths to be self-evident'?"

"The men just sit around all day," says Ariel.

"Waiting to get laid," says Rachel.

"Rach!"

"They are! They're relentless. They act all sweet but they say the nastiest things."

"They're like men everywhere," I suggest. "They're kidding."

"Of course they're kidding," says Ariel.

"They're half-kidding," says Rachel. "How do you say 'oasis,' as in, 'an oasis of freedom'?"

"Atchoni is the worst. I hate going there," says Ariel.

"Have you ever had any trouble?"

"Depends what you mean by trouble," says Ariel.

"How do you say 'sweltering with the heat of injustice'?" asks Rachel, swallowing her smile.

Soon an old woman appears in the doorway. Ariel runs and gives the woman a hug.

"*Abena-folo* is getting cut today," Rachel tells me. "You can watch, if Mama Iso will let you."

I follow the women outside. For the next forty-five minutes, Ariel scratches abstract pictures of the sun into the dirt with a twig. Mama Iso studies them meticulously. When Ariel is convinced Mama is on the same page she lifts the back of her own shirt, revealing a similar image already carved into the skin of her lower back. The tattoo looks raw and sore; this is her third cutting.

Mama Iso brandishes a razor blade, kneels down and sets to work reopening the wounds. Ariel grimaces in pain and holds Rachel's hand as Mama Iso digs the rays of the sun into her flesh. When I lean in to have a better look, Mama stops and covers her work with her hand. She says something I don't understand and then bursts out laughing.

"You'd better back off," says Rachel. "Or else."

"Or else what?"

"You'll get an erection."

These are *kammbá* tattoos, the same family of skin cicatrizations I saw on the whiskered face of the old woman in Balingsoela. For the Saramaka, *kammbá* are visual and tactile aphrodisiacs, heightening sexual desire in men and making women irresistible. Intended for the sole enjoyment of a woman's husband, *kammbá* have been a part of Maroon culture for centuries. The custom is slowly dying out, but many of the older woman here are said to have them.

"So, who is your tattoo for?" I ask Ariel.

"Nobody."

"I know!" says Rachel.

"Rach, don't."

"You know your boatman?" Rachel asks.

"Rach!"

"You mean Denny?"

"The tattoo is for me," says Ariel. "That's the whole point. It's not for Denny. Denny's a jerk."

"I thought he was a nice guy," I say.

"He is nice," says Rachel.

"He's also a jerk."

I have stumbled into a cross-cultural soap opera. It's the perfect situational drama, really, full of conflict, emotional and otherwise. Sex and self-immolation deep within Klaus Kinski's erotic jungle. I half expect a camera crew to step out from behind Ariel's Durotank.

"And anyway, *Badja-mai*," says Ariel. "You're one to talk. At least I'm not sleeping with a married woman."

Lights, camera, action.

Rachel squeezes Ariel's hand hard and gives me a guilty look. Ariel yelps in pain as Mama Iso's razor cuts deep. The waist of her *pangi* is soaked with blood. Mama Iso is oblivious to the revelations going on around her.

Homosexuality is anathema to Maroon culture. It is difficult to get the Saramaka to imagine a same-sex couple, let alone admit they exist. The denial is virulent among the men I've talked to, but I've never asked a woman about it.

"It's not like it's a secret," says Rachel. "Her husband knows."

"Her husband knows?"

"Sometimes he joins in."

And here we reach the outer limits of my naïveté.

"Look. We're still getting our heads around it all," says Rachel. "Me especially."

"We spent the first year here fending off the men," says Ariel. "We learned the culture and the language. Then we just got tired of the double standard."

"We got tired of drawing lines."

"The lines don't exist. Isn't that the point of Peace Corps? How do you draw a line between cultural and emotional exchange?

"And it's not like we're sleeping around. We're having real relationships. They're complicated and frustrating, but they're real."

I ask the girls what Peace Corps says about relationships with locals.

"Nothing," says Rachel. "There's nothing they can do. We're all alone out here for two years."

"We're young and in our prime!"

Mama Iso finishes cutting and wipes up the blood. Then she rubs a sticky concoction of black ash into Ariel's wounds with *adooya* leaves. The ash serves as the dye but also as an irritant, making the scars more prominent and therefore more erotic to the touch. When she's finished, Mama Iso stands up, gives a hearty laugh, and shakes her rump.

We eat fajitas for dinner. Then we go straight to church.

The pews are almost full when we arrive. Some of the villagers are wearing their Sunday best – clean T-shirts for the men, bright *pangis* and hair-wraps for the women. The service begins with music, led by Sheila on the ukulele. The air fills with joyous voices and plenty of clapping. Walter sits in front of me and sings the loudest. Then Wout steps to the front of the room, accompanied by the pastor of Abenaston. On the wall behind them, a makeshift sign paraphrases Psalm 107: *Praize Masra bikasi hem so boen.*

Wout has chosen John 3 for tonight's sermon. He begins softly, humbly, speaking one verse in Dutch and then allowing the pastor to translate it into Saramaccan. He introduces Nicodemus, ruler of the Jews, and describes his teaching at the hand of Jesus. Nicodemus had no trouble believing Jesus was a man of God; his difficulty lay in the idea of spiritual rebirth. So he asked Jesus to explain how an old man can be reborn. "Surely," Nicodemus said, "he cannot enter a second time into his mother's womb." Wout becomes visibly anxious. He rocks from side to side and carefully eyes the pastor, who is no match for Wout's oratory and struggles to find the right words.

The pastor finishes and Wout launches into a vigorous narration of Jesus's response. His voice booms through the one-room church.

He waves his arms and stamps his feet, barely letting the pastor finish before hollering the next lines from his script. He talks of water and flesh and wind. He comes again to Nicodemus's disbelief and Jesus's condemnation of disbelievers. Then something small and dusty lands on my head: a termite nest lodged in the rafters seems to be crumbling under the force of Wout's evangelism.

Sweating profusely and breathing hard, Wout finishes with a rousing flourish. The pastor does his best to follow suit but ends up bursting into laughter. Sheila takes up her ukulele again and Walter begins to sing. The villagers are slow to join in this time; they are still processing the message of Wout's story. When the song ends, a middle-aged Saramaka man stands up and shouts something to his pastor.

"It seems we have a Saramaka Nicodemus in our midst!" yells Wout in English. The villager is also having trouble with the idea of emerging a second time from his mother's womb. His mother is dead, for one, and his father has two other wives in a village upriver. He wants to know if he can return to one of their wombs, and if so, how is he to choose?

Wout does his best to reframe the story in less metaphorical terms but the man still looks confused. Sheila starts playing again. The villagers stand and begin singing and clapping to the beat. Children leap through the pews. Candles flicker in a sunset breeze wafting in from the river. Wout beams and puts an arm around his pastor.

The church empties slowly. The villagers pause at the door to thank Wout and his friends, who are very gracious and thank the villagers in return for the hospitality. As we're leaving, Ariel says something to Wout as she's shaking his hand and he explodes.

"Impossible!" he yells. "There is no devil worship here!" The few remaining villagers head quickly for the exit.

"I agree with you," says Ariel. "But I've seen what I've seen."

"Impossible!"

"I've seen it."

Wout removes his glasses, closes his eyes, and pinches the top of his nose. "Thirty-three years ago zee people here were heathens," he says quietly, his face acquiring a rare rush of colour. "Zey walked around with metal links around zair necks. Around zair elbows, around zair wrists. Zey had multiple wives. Zey worshipped a sorcerer."

"He's called a medicine man," says Ariel.

"Yes. A sorcerer witch doctor."

"No. A medicine man. And just because people dress differently doesn't mean –"

"Enough!" yells Wout. "Zee people here are all good Christians! Zey worship zee Lord our Saviour!"

"With all due respect, sir, you don't live here," says Ariel.

"With all due respect, young lady, neither do you."

Ariel storms off. Rachel and I catch up to her outside.

"This guy's living a fantasy," says Ariel as we walk. "Sure, the people go to church. But everyone still believes in *winti*, the forest spirits, divination, the old religions. The ancestors are still here."

Back at Ariel's house, I collapse into my hammock. As I'm about to nod off I remember Martin Luther King.

"Hey, what about the King speech?"

The girls groan from their beds.

Early the next morning I meet Wout and Willem in front of the village schoolhouse. Walter and Sheila are leading the children in a rollicking recital of Christian songs. Between hymns, the couple hold up huge poster-boards inscribed with snippets of scripture and simple drawings of innocent, dark-skinned children doing wrong and making right. Each morality tale ends with a take-home message. The children struggle to repeat them in their stilted Dutch.

Wout claps his hands and Willem does nothing. After three stories, the principal appears on stage and ushers the children inside.

I join the missionaries beneath the huge mango tree at the shore. Denny directs us to his boat, which is sandwiched between two other dugouts. Scrawled on the hull of one of these crafts are the words *Neks No Fout* in bright blue paint.

Neks No Fout was Desi Bouterse's slogan during the last presidential election, when the old dictator entered the race intent on regaining the power he'd once enjoyed, this time in a more democratic fashion. The phrase means, essentially, "no harm done," and is seen by many as an arrogant, sideways reference to the tragedies of Suriname's recent past – tragedies Bouterse played a central role in and, in many cases, personally orchestrated. The slogan is a boorish insult to those citizens who are unable or unwilling to let those atrocities go unpunished.

The Upper Suriname, a tangled jungle accessible only by gruelling journeys, is still not too far from the city to harbour at least a small measure of that reliable urban signifier: biting political cynicism. Exploiting perhaps the only form of mockery available to a rainforest people, the Saramaka have responded to Bouterse's brashness with a measure of modest ridicule. Over the last two months I've seen at least ten dugouts nicknamed *Neks No Fout*.

With our gear stowed and a few villagers waving from the stones, Denny aims the boat upriver. Wout, Sheila, and Walter talk excitedly about the schoolchildren but Willem, as usual, is quiet. He just stares out over the water at the dense jungle around us. Fifteen minutes into the ride he leans forward and taps me on the shoulder. "Beautiful," he says softly, referring not to the evangelical works of his colleagues but instead to the divine works of nature. It occurs to me that the gospel of ecotourism and the "conservation economy" has come dangerously close to drowning out the truth, that there will always be deeper spiritual reasons for leaving nature be.

Soon we reach the rapids at Felulasi, where we almost tip as Denny powers up the steepest section. He yells for us to lean to one side but the missionaries think he's kidding. Our saviour is Willem, who leaps from his seat to right the boat. Sheila and Walter scream and grip each other tight. Wout doesn't move but his face turns a little more ashen.

At the top of the rapids we pass three villages in quick succession. Wout claims they are all good Christian communities. He continues listing off his triumphs as we motor south, a modern-day conquistador in a bright-red life jacket, but as we pass more and more sacred gateways and ancestral shrines at the water's edge, he falls silent.

For a short distance south of the Wild Coast, the rivers of Suriname are brackish and lined with thick groves of mangrove trees. Then the saltwater peters out and the forest of black thorns and giant mosquitoes gives way to the majestic Guianan jungle that blankets the rest of the country. We are now deep in Saramaka Maroon territory. Some villages are at various stages of conversion to Christianity but most are still, in Wout's words, "heathens lost to the wild." We see fewer churches and, Wout is quick to point out, "more women with their breasts hanging out." Like the mangrove swamps at the mouth of the Suriname River, even the culture-corroding power of Jesus and his crown of thorns has its southern limits.

The village of Futunákabá was founded in the early nineteenth century as a Christian mission in the heathen jungle. The place got its name from a speech the local pastor, Anaké, gave around the time of the town's inception. In describing the potential of the place to attract the unconverted, he apparently said "*futu ná kabá*," or "the feet will never stop."

Denny pulls to shore and I help the missionaries unload their gear. They make their way up the path but I hang back with Denny.

"Wout," I say. "Denny and I are going on."

Wout turns and walks slowly back to me. "What?"

"We're going farther. To Asidonhopo. Denny thinks we can make it by dark."

The others return and surround us. Sheila and Walter look worried. Wout looks angry. Willem is smiling and lost in thought.

"Don't be foolish," says Wout.

"I'll report back," I say.

"I have been south. Zair is nothing zair."

"There are many villages. Thousands of people."

"Zair is darkness. Believe. Zat is all."

"Father," says Sheila.

"He is a foolish boy! His boat has come to shore!"

Sheila puts her hand on Wout's shoulder. "He has made up his mind."

"We should pray," says Walter.

"Yes," says Sheila, reaching out her hands.

"Denny," says Walter. "Can you make it by dark?"

"*Ai.* No problem."

"You will need gasoline."

"*Ai.*"

"You must go fast."

"No problem."

Now Sheila, Walter, and Willem encircle us, holding hands. Wout reluctantly joins them and bows his head. He says a short prayer, asking the Lord to protect me from evil, to help me find what I'm looking for and return me to my father's arms. Then Wout leans forward and, without a word, pulls me in for a bear hug.

The missionaries gather on the rotten dock to see us off. Denny unhitches the boat. Then Willem shoulders his backpack and leaps awkwardly into the bow.

"I am coming with you!" he says.

"Brother Willem!" yells Wout.

"It's OK, Father. I want to see this darkness."

Sheila and Walter are speechless. Wout just shakes his head and starts up the hill.

We buy gasoline at the village of Pikiseei. When I tell the gas jockeys where we're headed they shake my hand and ask me to remember them to the Gaama. Soon we're back on the water, powering through a series of rapids, and Willem is telling me about his work.

"The prisons of Siberia are filled with sorry men," he says. "Putin wants them emptied, but you cannot simply release men like this. They are shattered people. Many times, it is hard to tell if it was life or the camps that broke them." Willem's television network spends millions of Euros every year sending missionaries to some of the most desperate corners of the globe.

"Rehabilitation through the Lord Our Saviour," he says. "We are the only source of hope for these people."

I ask if he's drawing a spiritual parallel between hardened criminals who've been banished to forced-labour camps in the Siberian tundra and the Maroons of Saramaka.

"Of course not," he says. "Our work here is much easier. These people already live in paradise." Willem smiles and begins rooting through his backpack. He retrieves two bananas and hands me one.

"You don't think we should be here, do you?" he asks.

"No."

"You think we are acting like conquerors."

"In a way."

"May I ask you something?"

"Sure."

"What brought you to Suriname?"

I'm not sure how to answer. Willem smiles again.

"I'll ask it another way. What were the signs that led you here?"

"There were many."

"Such as?"

I still don't know how to respond. (Emma asked me the same thing a month ago on the phone.) Willem waits for a long time. The motor howls as we reach another small falls. We pass the village of Daume, named after Dahomey, the ancient West African kingdom.

"You and I are very similar," he says finally.

"You think so?"

"Yes. You may not believe it, but I do not feel like a conqueror. Neither does Father Wout. In our hearts we are just like you."

"How's that?"

Willem tosses his banana peel overboard.

"A bunch of ignorant men following signs."

Although Richard Price's book tells the stories of many Saramaka clans – the Abaisa, Langu, Nasi, Dombi, and Awana – as well as tales involving the Ndyuka, Paramaka, and Matawai Maroon tribes, most of them concern Matjau beginnings.

After the Great Raid, Ayako and his fellow Matjau escaped deep into the jungle but continued to foment rebellion on the plantations. First-Time narratives describe how Ayako used Affima, his mighty war *obia*, his supernatural charm, to save the life of his son, whose mother hid him in the reeds before the Dutch military attacked their village. We learn how Ayako used another *obia* to free Gunguukusu and his people, the Watambii, thereby forging the fierce political alliance between the two clans that exists to this day. We discover how a benevolent spider monkey helped a woman named Zoe found the Watambii Twin Cult, which draws heavily on the folklore and myths of Africa. And we meet the Seven Who Walked Together, an allegory for the journey Ayako's rebels made from Matjau Creek south to the Upper Suriname region, which Richard Price calls the Great Southward Trek.

Price explains, "For Saramakas today, talk about First-Time is very far from being mere rhetoric, preserved for reasons of nostalgic pride. Rather, First-Time ideology lives in the minds of twentieth-century Saramaka men because it is relevant to their own life experience – it helps them make sense, on a daily basis, of the wider world in which they live."

Time slows as we push farther upriver. The villages look older and more weather worn. Sacred gateways proliferate and the rapids become more complex. Silk-cotton trees soar above the canopy, their branches filled with nests, their ancient trunks rippled with sun. As we pass one of these monsters, a four-foot iguana leaps from a branch and belly-flops into the river. Willem and I lean over the gunwale and watch the animal swimming beneath us, its rack of spines flexing to the rhythm of its tail.

In the late afternoon, Denny pulls to shore at a large village, where a few men have gathered on a small pebble beach. Denny tightropes past me along the gunwale and tells me to wait. We have reached the Djumu Mission, site of the only hospital in Saramaka.

Until 1980, says Richard Price, this hospital had never treated a single broken bone or gunshot wound, even though these injuries were common in the forest. Instead, injured Maroons would travel to the villages of Daume or Kapasikee, where ancient *obias* specific to these wounds are housed. For the most part, the same is true today. These *obias* are the same healing spirits that were smuggled into the jungle by the first rebel slaves.

Denny chats with the men for a while. He glances back at us as he speaks. When he returns he gives me an embarrassed look.

"We have problem."

"What's wrong?"

"You want see Gaama?"

"Yeah."

Denny sucks his teeth. "Sorry," he says. "Gaama gone."

I have been travelling in Suriname for nearly four months now. I've become used to disappointment, to unexpected sadness and difficult journeys. This is worse.

We drift in the middle of the river with the engine off. Denny and I face each other and discuss our options. While we talk, a yellow butterfly from a García Márquez novel sits on the brim of his hat.

"We can go back," he says.

"But we are already here."

"Yes. But we can go back."

I feel a familiar anger returning.

"How much farther is Asidonhopo?"

"Not far."

"Then let's go."

"Gaama not there. He go north today."

"Then I'll wait for him."

Denny shoos the butterfly. "You cannot wait for him."

"Why not?"

"Gaamakondre not for tourists."

We stare out over the water in silence. The butterfly returns and lands on my hat. The sun is dropping.

"We have come a long way," I say.

"Yes."

"If we go back, I waste a lot of money."

· "Not waste. Spend."

Now Willem speaks up. "What will happen if we go to Asidonhopo?"

"*Kapten* will tell us to leave."

"Then we will!" I say. "If he tells us to leave we will leave!"

Denny frowns and goes quiet again. Then he stands and walks past me along the gunwale.

"Ok," he says. "We go. Then we leave."

I look at Willem, who always seems to be smiling.

"It is not the end of the world," he says.

In five minutes we reach the headwaters of the Suriname River. To the right, the Gaanlio is a gorgeous field of waterfalls, the rapids of Tapawata. To the left, the smaller Pikilio winds away into the bush. Between the two is a narrow spit of jungle called Tuliobuka.

For all Saramaka, the most important First-Time story took place right here more than 250 years ago. The legend tells of a rafting expedition Ayako mounted on the Gaanlio, accompanied by his great *obiama* Kwemayon, the African shaman who was always at his side. The two poled their way up the Gaanlio to an island upstream from the falls of Tapawata, where Kwemayon performed divinations, asking the river for permission to live there.

As the elder Tebíni, Price's best source and teacher, tells it, "Then they continued, descending the great falls with their raft, on the side of the *kwama* reeds. There, at the confluence of the two rivers, Ayako 'cut' the reeds, claiming the Pikilio forever for Matjaus."

For Saramaka, this cutting of the reeds is the most significant human action recorded in First-Time, even more significant than the Great Raid on Cassewinica Creek. This story, including Ayako's subsequent wanderings through the region, is the basis for the Saramaka conviction that the entire watershed of the Suriname River is their home by divine right. Tuliobuka is the most hallowed site in Saramaka, where the Gaama still leads rituals in times of drought or flood, where Gansa, the Mother of the Waters, still holds ultimate authority.

After Ayako cut the reeds, Kwemayon divined their next move.

"It was then that Kwemayon descended into the river. He slept there, underwater at the foot of the falls. The African! He slept there at Tuliobuka underwater until he came out and said, 'Wherever you can find a suitable place, we can stay there.' . . . Divination told them, 'Only at Baakawata will you be able to hide people successfully.' And that's the way it finally came to pass."

Some time between 1720 and 1740, Ayako and the Matjau settled at Baakawata, "the place that could hide people," south of here on the upper Pikilio. For the next forty-odd years, Ayako ruled Saramaka as the unofficial tribal chief.

Denny veers the boat onto the Pikilio. Soon the jungle gives way to a collection of clapboard huts. We have reached Asidonhopo. We pull to shore, where three men watch our approach. Denny leaps past me and speaks to the men. His Saramaccan blurs into a rapid stream of song.

One of these men is the *kapten*. He shakes his head vigorously while Denny speaks. They talk for ten minutes, then Denny starts walking back to the boat with a satisfied look on his face. But then a woman appears among the men. She is older, perhaps sixty-five, and wears a plaid *pangi* with another wrapped around her waist. On her head she's balancing two plastic buckets filled with dishes.

When Denny sees this woman he immediately stops and turns to face her. I think I see him nod, almost bow. The woman speaks with the *kapten*, who tells her our story. Meanwhile, Denny gestures behind his back. He wants me to join them. As I do, the women gives me a thoughtful smile.

"*Disi Gaama frow*," whispers Denny. The Gaama's wife.

The Saramaka queen.

I nod-bow and wait for the captain to finish. When he does, the woman walks toward me. The dishes on her head rattle as she reaches for my hand. She shakes it and smiles. Her name is Sameni.

"*A no trobi*," she says. Then she walks away down the path.

I look at Denny. His eyes are wide.

"Go!" he says, when he realizes I don't understand. "Follow her! She says it's no trouble!"

I run after the queen.

18

—

WAITING FOR GAAMA

I follow Sameni to a modern, two-storey building. Compared with the huts around it, the building is massive, with a wooden staircase leading up to a porch that overlooks the darkening river. In front a pole flies the Surinamese flag. This is the home of the Saramaka Gaama.

Sameni lowers her buckets to the ground and introduces me to her family, the Aboikonis, who have appeared at the top of the stairs. Mione is her son, a handsome, thick-set man in his mid-twenties, with a round face and bald head. Melissa, her daughter, is a little younger and very beautiful, with a slender frame and tight dreads pulled back in a ponytail. Melissa's son, Jules, is three years old and looks mischievous. He has chubby cheeks and an enormous head. He is going to be huge when he grows up.

Denny, Willem, and the *kapten* arrive with my gear as Sameni unlocks a large wooden door beneath the porch. The door opens into a room the size of the whole house above. The windows are draped with embroidered fabric and laundry is drying on ropes strung from a central pillar. The floor is cement. A lone light bulb glows in the middle of the ceiling.

"You can live here," says Sameni, grabbing a broom and sweeping a pile of dead insects out into the night.

I look at my entourage and they burst into nervous laughter.

"*A no frede,*" says the *kapten.* "Don't be scared."

"*A frede?*" asks Sameni, grabbing my shoulder.

"No, *no frede*," I say. "*Soso breiti.*" I am happy.

Denny and Willem drop my bags and wish me luck. They have a long, dark ride ahead of them. Both of them shake Sameni's hand. I thank Denny for his patience and pay him his fare. He seems a little dazed.

"You have been blessed," says Willem.

"Make sure to tell Father Wout."

"I will," he says. "If he's still speaking to me."

The Saramaka royal family – the queen, the princess, and the two princes – leave the room and pull the door behind them. I hang my hammock among the laundry, open a can of tuna, and crouch on the floor to eat. Evening insects begin to hum and invisible rodents scurry across the rafters. From the village across the river comes a volley of fifteen gunshots, followed by raucous drumming. Then someone knocks on my door.

Sameni is bent over, her shoulders and arms bulging with effort, balancing a large wooden table on her back.

"They celebrate," she says, referring to the rifles. "Last week, big man died. Tonight is last night of mourning."

I help Sameni carry the table into my room. She motions toward a window.

"*Puti drape.* You will have light in the morning."

"For eating," I say. "Thank you."

Sameni shakes her head. "Not for eating," she says. "For writing."

The Gaama's secretary, a man named Samson, bursts into my room at seven the next morning. He is followed inside by a stream of villagers who have come to sign the work register. "Come," he says, scribbling his name quickly. "We go office."

We soon arrive at a circle of beautiful A-frame houses, smaller than the Aboikonis' home but much larger than the one-room huts we passed on the way. Samson leads me into the largest one, the Gaama's "residence," the Saramaka White House. I remove my shoes and step over an open bottle of beer on the porch. "Celebrate New Year," says Samson, pointing to the *jugo*. The bottle has been there for three weeks.

A long and windowless room. The familiar hiss and garble of the radio, or *sender*. A black office desk strewn with papers. A line of empty chairs against the wall. Carved rafters. An ancient rotary telephone on a rickety table. An officious air lent weight by Samson, who is already scribbling in a notebook.

"First, your name."

"Andrew."

"American?"

"Canadian."

"Ah." Samson nods. "You come yesterday."

"Yes."

"With who?"

I list off their names.

"And you are journalist."

"Not really. Writer."

Samson looks up. "Same thing?"

"Maybe," I say.

He nods. "Kindly, what is the purpose of your visit to Gaama-kondre?"

"*Koiri.*"

Samson looks up. Maybe *koiri* isn't specific enough.

"I would like to ask the Gaama something."

"Journalist?"

"No."

"Gaama *gwe* yesterday."

"Do you know when he's coming back?"

Samson shakes his head. "We talk every day. You want wait?"

"If it's OK, I wait."

"Good. Gaama will be very happy." Samson points to the far wall without looking up. "Gaama likes visitors."

The wall is covered with framed photographs, some of them very old. Each is a photo of the Gaama – I count at least four different men, each a successor in the Aboikoni line – surrounded by Dutch royals, influential politicians, and famous Americans. Queens Wilhelmina, Juliana, and Beatrix of the Netherlands; past and present Surinamese presidents; well-known Caribbean anthropologists; Russell Mittermeier, co-founder of Conservation International; a few smiling Peace Corps volunteers.

"Where did the Gaama go?" I ask.

"He go Brownsweg. Then Rosebel."

"Is there *jugo-jugo*? Trouble?"

"No." Samson smiles falsely and shuts his notebook. "No *jugo-jugo*. Meetings."

"With who?"

"Canadians. Like you!"

"Cambior?"

"Yes."

"Something wrong?"

"No, no, no. Some of our people. They were caught."

"At Rosebel?"

"Yes."

"Doing what?"

"Digging for gold."

A voice crackles over the *sender*. Samson leans in to listen.

"Sounds like *jugo-jugo* to me," I say.

The room goes dark. In the doorway stands an enormously obese man. He wears green athletic shorts that barely begin to cover his massive thighs. The rolls of fat on his torso are so thick they look like rolls of muscle. The man nods to Samson. Then he sits down next to me with a massive grunt and reaches out his hand.

"Basia Joe."

Soon we are joined by three more men, all of them *basias*. For the next fifteen minutes, I listen to the men talk over each other in a rapid stream of Saramaccan. They all seem angry. Perhaps they are speaking about the incident at Rosebel.

Basia Joe looks at me and frowns. "*Taki-taki*," he says to the others. He suggests they switch to Sranantongo so I can understand what they are saying.

The men are not talking about Rosebel. They are discussing a villager who claims the Gaama owes him money. The *basias* are split on the topic – two believe the villager should be paid, two believe the villager hasn't worked for the Gaama in over a year. Each side accuses the other of pandering to friends. Tempers flare and Basia Joe heaves with fury. The only employment opportunity in Asidonhopo – one of the few paid jobs in the Upper Suriname Region – is at the Gaama's political office as either a *kabiteni, basia,* or low-level aide. This work is fiercely sought and strictly monitored.

The phone rings. Samson picks it up and shields his mouth as he speaks. The *basias* continue to argue. Then Melissa appears at the door and everyone falls silent.

"*Andu?*" she says, her smile dazzling. "*Wi gwe na goón.*" Melissa and her mother are leaving for their *goón*, their garden plot.

I look at the *basias*, who return my gaze sheepishly.

"*Oten?*" I ask. When?

"*Dalek*," she says.

"*Mi wani kon.*"

Basia Joe snorts. "*San? Man wani go na goón!*" The other *basias* snicker. It is not manly to visit the gardens.

Melissa thinks for a moment. "*Yu wani?*"

"*Ai.*"

She smiles again. "*Kon gwe.*"

I follow Sameni, Melissa, and Jules along a slim path that leads to Dangogo, the southernmost village on the Pikilio. Between the two villages the forest is dark, impenetrable at the side of the trail. Occasionally, though, the vegetation is interrupted by bright clearings, slash and burn garden plots where the women grow their staple foods. We pass a number of these sun-drenched patches and wave to the gardeners, women bent double with effort. The *goón* is the only place beyond the cooking fire where a Saramaka woman has real authority. It is a refuge, a small parcel of land carved from the bush where women work, socialize, gossip, and get away from the men.

The clearing of a new garden is, however, a community affair – the men clear the underlying brush, chainsaw the lianas and trees and chop them into firewood. Then the plot is set on fire. The villagers celebrate with a feast, thanking the jungle and each other for their hard work – from here on, the land is the sole responsibility of the women. Some families – the ones with high social standing – have more than one *goón*, each planted at a different time, ensuring a continual supply of vegetables.

In half an hour we turn off the main trail and hike along a faint track. I walk behind Sameni, who carries a bucketful of gear on her head, an overstuffed backpack on her shoulder, and a machete in her hand. Melissa is behind me, weighed down with a bucket on her head and Jules on her back. Jules has been crying most of the way. He does not like the jungle, Melissa says, and hates to walk.

I ask Melissa if she's ever seen a jaguar on these trails. She sings a questioning, "*San?*" and then laughs. No, she says, she has never seen a jaguar here. But many people have. When she was young, her mother told her never to walk alone in the jungle on Tuesdays.

Tuesdays are Jaguar days. This was enough to keep her out of the bush every day of the week unless accompanied by a group of women.

We cross a small stream and reach the Aboikoni gardens. The plot is a two-acre swath of scorched earth and blackened stumps, but life is already bursting from the ruins. Green leaves the length of my arm splay sunward, perched on stalks the width of my wrist. Gold-brown cassava tubers angle up from the earth like giant, swollen fingers. Trains of delicate squash meander among decaying branches, and a stand of sugar cane glistens like bamboo. The abundance is stunning, an edible jungle grown from scratch. And although I know it won't last, and the ash-enriched soil will be exhausted before too long, it is hard to imagine a better illustration of fertility – of resurrection – than the green from black burgeoning of a Saramaka garden.

Sameni opens her backpack and pulls out a rusted tank and a long hose. Then she ties two plastic bags around her hands and walks with the tank through her garden. Pointing the nozzle at unwanted weeds, Sameni sprays a faint herbicide mist throughout the plot. Her feet are bare and she doesn't wear a mask; at times I can barely see her through the haze. This is modern Saramaka gardening. When she returns an hour later with beads of what I hope is sweat on her brow, she wonders aloud if the chemicals in the tank might kill her, too.

We stay at the garden all afternoon. Sameni tends to her cassava, green beans, and *angroki* while Melissa and I entertain Jules. The boy picks through my backpack and puts everything he can find in his mouth – pens, notebooks, digital flashcards. He is relentless. When the rains come the women take shelter under a tarp, but Jules refuses to leave my side. We cower together beneath a banana frond, the young prince untying my shoelaces and chewing on the tassels of my jacket.

We leave at sundown. Sameni walks with an armload of firewood and another pile stacked on her head, Melissa carries the gear, and I give Jules a ride. The little man quickly falls asleep on my shoulder. By the time we reach the village I am almost deaf from his snot-filled snoring.

I cook dinner for the family with food I brought with me – penne and tomato sauce with tuna and hot peppers – and Sameni marvels at the sight of a man working in her kitchen. Basia Joe comes by, right on time, and helps himself to a plate. Meanwhile, a cry goes up from Akisiamau on the other side of the river. Someone has shot a wild boar, and the villagers are jostling at the shore to buy meat. Fresh meat is rare around here; hunters have to travel upriver for days to find good game.

At Sameni's suggestion, Melissa digs out the family photo albums. While her mother shells peanuts, Melissa describes each photo, recounting everyone's names and their kinships. There are pictures of the Gaama during his official swearing-in ceremony, dressed in his traditional outfit of cape and *kamisa*. In one of them he is weeping, obviously overwhelmed, with Sameni sitting stoically at his side. Mione appears in the doorway with a shank of fresh boar, his forearms covered in purple blood. He dumps the meat in the sink, leaving it for his mother to deal with.

Sameni says nothing until we reach the last photo. Then she nods her head.

"This is your family now," she says to me, pointing at her children. Melissa and Mione laugh, but Sameni claps her hands to quiet them. "Mione, your brother. Melissa, your sister. Me, your mama." Now she laughs. "*Yu mus wakti.*" I must be patient. The Gaama will be worth the wait.

Soon the power goes out, the kids are off to bed, and it's just me and the queen, my Saramaka mother. We sit in the dark for a few minutes, the only sound the rhythmic snap and crack of peanut shells. Then I flip on my flashlight and point it at Sameni's hands.

She knows what I'm thinking – that she hasn't stopped working since I arrived. She giggles.

"I want to sleep but the work won't let me," she says, almost crying with laughter at the story of her life. "The work won't let me sleep!"

The next morning, Melissa, Jules, and I paddle across to Akisiamau with the gardening gear. We hike through the village and into the bush, Jules again complaining the whole way. I eventually give him a lollipop and sling him over my shoulder. Over rickety bridges and past bustling garden plots, we arrive at Vliegveld Djomoe, the Djumu airstrip, where Mione mans the *sender*. All is quiet on the airwaves this morning. To pass the time, Mione pecks at a small electronic keyboard, composing lyrics over an amateurish tune.

The family has a second garden plot at the north end of the runway, and here we set to work. Armed with short garden hoes, Melissa and I turn the black soil and pile it into long mounds. These will be the peanut fields.

We take breaks to suck mangoes and watch Jules stumble around in my sandals. We hit a series of ant nests, their vicious bites sending us leaping. In two hours my hands are dotted with white blisters and I am drenched in sweat. And then the rustle of Tapawata Falls reaches us on the wind.

From here, I can see the land narrowing where the Gaanlio and Pikilio converge. We are standing on Tuliobuka, the most sacred spit of land in Saramaka, now sullied by a government airstrip. This is where, three hundred years ago, Ayako cut the reeds. I am planting peanuts in hallowed soil.

"*Mi skin ati*," says Melissa. Hard labour like this makes her *skin*, her body, hurt. She prefers Paramaribo, where she attends school and works nights at Amigo Universe, an upscale Chinese restaurant. No one is buying peanuts any more, she says. She doesn't know why her mother bothers to grow them. I try to imagine the Saramaka princess bussing tables, while Melissa tells me her real dream is to guide tourists in the bush.

Sameni arrives at noon. When she sees the work we've done she pulls me in for a muscular hug.

"God has sent you," she says dramatically. "If it wasn't for you I'd be dead!"

We eat lunch with Mione. Boar stew with rice and *bitawiri*, a bitter leaf boiled to a mush. The meat is rich and gamey and Sameni makes me eat three helpings. Then we go back to work. The afternoon passes in a hot, hazy daze. My hamstrings protest the constant bending and hoeing. Two bush planes pass over but do not land. As they approach, I hear a burst of voices from the *sender* and Mione howling back. Sameni disappears for three hours to help another women work her land. She has been blessed by my presence, she says, and wants to spread the wealth.

The rains threaten but do not come, and by five o'clock our work on the *goón* is finished. As I wash up, a pair of women arrive with their young children. One of the girls, no older than two, screams when she sees me and runs to hide behind her mother's skirts. I am the first white man she has ever seen.

We walk back to the river as the sun sets. At the Akisiamau *winkel* we find a gang of men – teenagers, the elderly, *kaptens*, and *basias* – drinking *palum* and watching a Papa Touwtjie DVD. It's a video of the famous concert he gave last New Year's Eve, six months before he was shot for the second time and killed. The rapper wears a black vest over a sharp dress shirt and a black-rimmed hat over a white doo-rag. He is the epitome of Surinamese cool. The women paddle back to Asidon with Jules while Mione and I stay for a drink.

The men are members of the Watambii clan, fierce allies of the Matjau, descended from those slaves Ayako helped liberate in First-Time. But while it is agreed that their ancestors "walked behind" Ayako and his people, the Watambii version of their freedom story also tells how their people brought fire to the Matjau, repaying their debt to Ayako and claiming status as proud equals. In this way, First-Time ideology serves as both a buttress to modern-day alliances in Saramaka and a tool by which history can be selectively, and flatteringly, cast. Matjau stories make no mention of having been given fire.

The men have obviously been boozing all day. The young are dressed as if waiting for a boat to Paramaribo, fresh white T-shirts with hip-hop logos, shorts past their knees, plenty of fake bling around their necks. The older men look haggard, with long dirty T-shirts and long, tired faces. Young and old alike exude a vaguely threatening, unpredictable machismo that would make me wary were I not accompanied by the Gaama's son. Some of them lurch drunkenly to the music. Others nod icily when I catch their gaze. I could be standing among the black market dealers on Saramaccastraat, aside from the trio of macaws screaming overhead as they return home to roost, the burn of fire ant bites rising up my legs, the hum of a wasp nest somewhere behind the store. These men are without occupation, without anything to keep them busy – job opportunities on the coast are a long way away. Surely the older ones can remember a time when their culture and the surrounding jungle provided everything they needed – an economy, an identity. Now, it seems, the young remember nothing and those times are over.

Or perhaps I'm wrong.

Because the stories of First-Time are relived every time a man leaves his village for work on the coast, every time a vote takes place in Suriname's parliament, every time a foreign company gains title to a new tract of Saramaka land. These young men may not know who Ayako or Kwemayon were, or what they accomplished. But those ancient stories are the ground beneath the people of Saramaka.

We stay until nightfall and paddle back home. I cook another meal, masala mango chickpeas with fried onions, and wander the village by flashlight with a plate piled high. I find Basia Joe outside his hut. His enormous body is wrapped in a puny beach towel and he is surrounded by children. He thanks me for the food and claps me on the back, then squeezes his bulk through his tiny doorway.

I meet Matthais Mitchell, the man who shot the wild boar. I've seen him wandering the village, a slim man dressed entirely in khaki with a rifle slung over his shoulder and a belt of *patrons*, shotgun shells, around his waist. Matthais is the best hunter in Asidonhopo. Tomorrow afternoon he will go to Pineapple Mountain. I am welcome to join him, he says. He will leave from the *winkel* at one o'clock.

At the White House, I find Samson slouched at his desk. A dim lamp casts long shadows through the room.

"Any news from the Gaama?"

Samson peers at the silent *sender*. "Noti."

"*Yu no taki?*"

"I wait all day."

Samson's notebook is open but I don't see his pen. In Saramaccan, *asidonhopo* means "sit down and hope."

"They no call," he says again, rubbing his eyes with his fists. "I wait all day for *noti*."

Ayako, first chief of the Matjau, died sometime in the late 1750s. His death marked the end of the Age of Heroes, the first generation of escaped slaves, those who had been born in Africa and sold into the New World. From here on, Saramaka leaders were forest born, African by blood but South American by birthright.

Just a few years after Ayako's death and after nearly a century of war, peace finally came to Saramaka. Though First-Time stories of the various clans differ in the details, they all seem to agree that a Langu leader named Wii, through a combination of troublemaking and happenstance, "brought the peace" to Saramaka. Fleeing the Matjau after being shot by Ayako's son during the great man's funeral, Wii arrived in the neighbouring Marowijne Region just as the Dutch were conducting final peace negotiations with the Ndyuka Maroons. Wii, speaking with questionable authority, proceeded to tell the Dutch that Saramaka also wanted peace.

In September of 1762, a delegation of Saramaka led by the new Matjau chief, Abini, signed a comprehensive peace treaty with the Dutch at Sara Kreek. This agreement ensured an end to the bloodshed, full autonomy for the rebels, and, most crucially, legal possession of all Saramaka lands for the Maroons and their descendants. And here we learn the depressing truth that, in Saramaka, there is no limit to the cruelty of irony. Sara Kreek, the place where the Maroons' right to their lands was officially recognized, now lies beneath the poisoned water of Brokopondo Lake.

During the first decades of peace, the Matjau gradually moved back north to the lower Pikilio, where the villages of Dangogo and Asidonhopo lie near the sacred land of Tuliobuka. These years marked both the end of First-Time as lived experience and its birth as unifying myth, as the old stories began their journey downriver and through the generations, all the way to the old men Richard Price studied in the 1970s and the few who remember them today.

In the village of Dangogo lies the Awonenge shrine. Forbidden to outsiders, black or white, Awonenge is the most sacred shrine in Saramaka, a tribute to Ayako, Kwemayon, and all the First-Time people who "heard the guns of war." It is here where libations of sugar-cane beer are poured, where new flags are raised, where the old songs and dances are performed, and where the names of Saramaka's heroes are invoked to the beat of the *apinti* drum. And it is here, Price says, where words first spoken during the peace talks at Sentea are ritually recounted.

"Those people who didn't live to see the Peace, they must not be jealous. Their hearts must not be angry. There is no help for it. When the time is right, we shall get still more freedom. Let them not look at what they have missed. Let us and them be on one side together, those First-Time people! It is to them we are speaking."

Sameni wakes me at seven o'clock.

"*Wroko kiri mi*," she jokes, explaining why she slept in. The *wroko*, the work, is killing her.

Melissa sits at the shore with a slim fishing rod. Every minute or so she jerks her wrist and pulls a small *witi-fisi*, white fish, from the water. She lets me try but I am useless – the nibble of a two-inch fish is, to me, imperceptible. I return the rod and soon her bucket is full. We eat fried fish and cassava for breakfast and then paddle back to Akisiamau.

Jules is in a strange mood as I carry him to the airport *goón*. He is frightened, seized by terrified screaming and uncontrollable crying fits. Melissa just laughs, but I can feel the fear in his body as he grips tight on my back. His little heart is thumping.

We spend the morning planting peanuts. My hands are soon cramped from prodding holes in the thick soil and my skin has begun to itch. There's a fungal rash on my belly, my ankles are infested with chiggers, and my legs ache. To make matters worse, the rhythm of bending and planting has worked its way into my head somehow, transporting me back a few months to the rhythmic and haunting dreams I suffered as I explored the lake and weaved through the bones of this river. With each peanut I bury, I try to bury Emma. And the more I remember her, the more fiercely she takes root. Like a seed she grows, shooting from this sacred soil, our young-leaf love, my old-wood anger. She couldn't wait for me, and I couldn't leave this place, so now we've become this wilderness, she and I, this Guianan green and detritus brown, all that grows and all that dies. My every thought is suddenly shaded though the sun pounds down. Those dreams, that phone call, that sadness, that loss – everything I thought I'd forgotten.

Mi skin ati.

By noon the field is planted and we head back to Asidon. At the *winkel*, I find five men half-asleep, lounging on benches beneath a mango tree.

"*Mi luku* Matthais Mitchell," I say.

The men look up. One of them, the oldest, gives me a tired smile.

"Matthais *no dya*."

"He go *sutu*?" Has he gone hunting?

"No," says the man. "Matthais has no gasoline."

I sit with them. They've heard of me, the white man waiting for Gaama. I tell them about Pineapple Mountain and one of the younger ones suddenly sits up.

"You shouldn't leave the village," he says, passing me a fat joint and a lighter. "*Furu fufuruman. Den kisi yu.*" There are too many thieves in the area.

I buy a warm *jugo* from the *winkel*. It is half-price because it is more than a year old. I return with six glasses.

"No one ever buys us beer," says the boy.

"I am thirsty," I say.

"So are we," says a man named Albert.

"When you go back to America, you should buy me a motor," says another. "Then we can build a house together."

"Why do we need a motor to build a house?" I ask.

The man frowns and closes his eyes.

To the right of the store, a young girl dribbles a basketball between potholes in the cement. She shoots at an old rusted basket hanging from a branch and gets it in every time. Samson walks by. When he sees me he just shakes his head and keeps walking.

"What do you think about all day?" I ask Albert after a long silence.

"*Sa?*"

"In your mind. What do you think?"

Albert pulls a pair of tweezers from his shirt pocket. "*Moni.*"

"How to make money," says the boy.

"How to spend money," says the sleeping man. He is blitzed. We're all blitzed. Everyone laughs.

Silence. The sun disappears. The splash of children.

"Do you think backwards or forwards?"

Albert stares at me.

"I mean, do you mostly remember things or mostly forget them?"

"*Frigiti,*" says the boy.

"*Memre,*" says the man.

"We live," says Albert. "*Libi, libi, libi.*"

The girl is perfect from three-point range. We watch her absently, as if on television. The river drifts past. A mango thumps to the ground. My eyes begin to close.

"*Ganja* stops thinking," says Albert, passing me the tweezers.

I wake to hot rains. The men are gone. The joint is a smudge on the ground. The girl sits on her ball, legs splayed, staring at the molten sky.

I find Mione in my room, watching a movie on a portable DVD player.

"*Yu wani go* Pineapple Mountain?" he asks.

"Yes!"

"*Wi go tamara.*"

"*Oten?*"

"*Fruku, fruku.* Leave early. Come back night."

"Yes! *Mi wani.*"

"Mione take care of brother," he says proudly.

Before bed, Basia Joe takes me down to the river. "I show you *bigi fisi,*" he says.

A group of men is standing knee deep in the moonlit water, gathered in a tight circle, talking excitedly and playfully shoving each other. Soon I hear the fizzle of fire and the men abruptly scatter, leaving a lone man in the water, the hunter Matthais Mitchell, holding a lit firecracker. A smile flickers on his lips. He throws the bomb out into the current and covers his ears.

The explosion is deafening. A geyser rises from the surface and the river rains down. The men begin to yell. A few sprint back into the water while others run along the shore. They all bend over and peer into the murk. A few trawl sticks through the shallows. Others sweep their feet back and forth.

Soon the men go silent. They stare blankly at the water, waiting for something dead or something remembered to surface from the deep. My ears ring. Basia Joe lets loose a disappointed grunt. A chorus of frogs begins to bawl.

No fish.

Nothing.

I wonder how many of the old stories are still here, and how many more would have been lost had Richard and Sally Price not fallen in love with Saramaka in the 1960s – I am not, of course, permitted to ask. But by 2001, all but two of the men who initiated Richard into the intimate world of First-Time had passed to the world of the ancestors.

Jules is screaming. His howls are torrential, mixing with the rains and old sadness that arrived in the night. Upstairs, I find the boy and his mother cuddling on the porch, staring despondently out over the river.

The storm rages for most of the morning and then lets up, but the sky remains dark. I find Mione next door, huddled with friends on the stoop of the empty visitors' residence, where VIPs from abroad hang their hammocks.

"When are we leaving?"

Mione passes me a joint but I decline. "*Alen hebi*," he says. It's raining hard.

"But it's stopping," I say.

He peers up at the sky. "It will come back."

"So we will get wet."

Mione's friends smile. "*Blaka man no lobi alen*," says one of them. Black men do not like the rain.

"*Ma* white man *lobi*."

They scowl.

"We should go," I plead.

"Man, *alen hebi*!" yells Mione.

"But it's stopping!"

"When it rains we do nothing," say the others.

I offer to pay, to hire them as tour guides. Again, Mione refuses.

"Pineapple Mountain is just a mountain," he says.

An ugly anger rises in my throat – the anger of boredom, the wrath of a stuck traveller. I remember Seyfu, Bodi, how so many of my friends in Suriname seem to be caught in an endless cycle of waiting, hoping, forgetting.

I swallow my anger, reach for the joint, and walk back to the house. I find Jules in my room. He is naked and covered head to toe in peanut butter. Even the boy's tiny penis is flecked with peanut. The jar sits open on my desk, where my notes and books are all streaked with brown. My pack sits upended on the floor, batteries and medical supplies and toilet paper spilling from every opening. Thunder rolls overhead and Jules starts to scream again. He knows he shouldn't be here.

I grab Jules by the shoulders and shake him. *Out!* I yell. *Out!* He screams louder now, tears streaming down his cheeks. He tries to get away but I don't let go. I squeeze his tiny shoulders. I shake harder, desperate to get it all out of me. I want to turn this little prince upside down and shake his body until the answers fall out. What am I doing here? What am I looking for? Why couldn't Emma wait for me?

I march the boy to the door. He manages to pin his arms against the doorframe for a split second, but I shove him outside with such force he falls to the concrete. He is three years old. I carry him upstairs, the boy kicking me and hollering his family name over and over – Aboikoni! Aboikoni! Aboikoni! – as if to say, "Don't you know who I am?"

I deliver him to Melissa, who is staring out over the river. She apologizes and so do I. She takes Jules down to the river to wash as another storm begins. I walk south through the village and hike into the bush.

The jungle is dark and smells like rot. Occasional lightning shocks the underbrush to green as the rain pounds down. I walk, ashamed, slipping through the mud in my sandals and almost pitching myself off a small bridge. I pass the gardens and see no one. I pass the turnoff for the Aboikoni *goón* and keep going. The claustrophobia of the rainforest is doubled by the downpour. Between the thunder, all I can hear is the rush of rain and my heavy breathing.

I don't know where Pineapple Mountain is but maybe I'll find it if I keep walking. Maybe I'll walk all the way to Baakawata, "the place that could hide people," or swim across to Dangogo and sneak a forbidden peek at the Awonenge shrine. Maybe I'll find a vitality there, the living, vibrant culture I've read so much about.

I can barely see through the rain. The jungle cracks and booms on either side of the trail. I hear a massive tree fall, a stone's throw away, the early snap and sweeping plummet, the whoosh of a natural clear-cut. Then the sounds converge and visibility drops to zero. Fear arrives, the same fear I felt on the highway as we passed the car crash at Moiwana, a bone-level chill that takes me out of myself. This might be Tuesday, Jaguar day. I fumble in my pocket for my puny knife.

I study the storm. I hear something like cautious footsteps. I scan the teeming bush, push the safety on my knife, do what little preparing a naive, navel-gazing tourist can do. Then, from behind me, a gruff breath. A weight on my shoulder. I almost leap out of my skin.

The man is smiling.

"*Desolé*," he says, mistaking me for a Frenchman. He spins a tattered blue umbrella to the side as he passes.

When Richard and Sally Price first arrived in Dangogo in the mid-1960s, only one aspect of Saramaka society was forbidden to them. They were permitted to study the language, religion, artwork, and social structure of their hosts, but not the ancient stories. At the heart of First-Time ideology is the belief that First-Time kills, and although the pair didn't know it at the time, their presence posed a vast spiritual danger to their hosts. Many locals believed the Prices had come to destroy their world. During those early years, Captain Kala, a Dangogo elder, would routinely visit the Awonenge shrine to pray.

"White folks have never come to Dangogo. The ancestors always said whites must never come as far as Dangogo. No outsider [black or white] has ever slept in Dangogo. The Old-Time people simply cannot 'see' white folks. The war we fought, it's not finished yet. . . . What in the world are we to do with these people? I have never buried a white person. If they die, how will I know how to bury them?"

For those first two years of fieldwork, Richard spent most of his time hunting, attending oracle sessions, and participating in Saramaka life much as any young man should. "The Saramaka historian-in-training hoes a very long row," he writes, and those early years were necessary for a white man, or *bakaa*, to build familiarity and trust with the Matjau. But when he returned in the late 1970s, things had changed on the Upper Suriname. The flooding of Brokopondo was complete. The Western world had seeped in, in the form of occasional tourists and film crews. Government officials visited almost monthly. Some men now wore long pants instead of breechcloths and spent much of their time earning money on the coast. And Price had also changed. He was no longer a student but chairman of an academic department and considered an authority on Saramaka culture. The people of Dangogo now welcomed him as someone who might be able to help them. Knowledge is power, says the First-Time historian, and Price was about to learn the truth of this.

He spent two field seasons working exclusively with the elders of Saramaka, discussing and recording their stories. He worked mostly in the evening, but also attended numerous sessions at "cock's crow," before the village had woken, where he listened as the men schooled selected youths in the tales of First-Time. Price became skilled in the interview arts – patience, tact, and strategy – as he discovered another First-Time truth, "never tell another more than half you know." Beyond the general taboos, the ancient stories were further protected by the storytellers themselves, who employed a frustrating and playful combination of interruption, digression, and evasion. Even the names of the main protagonists were changed, since to speak the name of an Old-Time person would ensure that the terrible, old days would return. Ayako, the first chief of the Matjau, was not Ayako – he was actually Lanu, a man whose name is rarely spoken. Ayako was Lanu's brother.

Slowly, Price learned enough about his teachers to interpret their tales. Between field seasons he sifted through the archives of the Algemeen Rijksarchief in the Netherlands, piecing together the arcane story fragments with obscure Dutch colonial documents. His understanding deepened to the point where he was able to conduct profound conversations with his teachers on the themes and metaphors contained within First-Time. This was a gruelling process, involving in some cases more than twelve years of building relationships, and his sessions with clans other than the Matjau were often riven with mistrust and fear.

In the end, Price's perseverance paid off. When he arrived back in Dangogo in 1978 he was greeted by Captain Kala, who had been, he says, his "fiercest adversary during the often-difficult first two years, and a strict upholder of the First-Time ways." Kala greeted him with a proverb, drawing parallels between the Prices' involvement in Saramaka and the eventual blossoming of a new garden.

"When it's dry season and you begin to make a garden, you risk death at every turn. When you clear the underbrush, your machete can kill you, a snake can bite you, a tarantula can bite you; every sort of thing can kill you when you're clearing the underbrush! [Then] when you go to fell a tree, well, every single tree can kill you; the axe in your very hand can kill you. You do all those things [take all those risks] right through the time when you burn the garden [the final stage before planting]. And then the rice grows."

For Price the real reward of his research had little to do with the stories he slowly learned; instead, he was more moved by the exasperating and perplexing process by which he gleaned his sacred education from the Saramaka elders.

"As I think back on my more difficult encounters or listen again to the tapes that are so frustrating if viewed solely from the perspective of gathering facts, I am struck by the overwhelming dignity of these quiet elderly men. If I learned less from them than I would have liked about First-Time, I learned from them something far more important."

The elders' commitment to the rituals of evasion convinced Price of what he had always suspected, that the stories of First-Time resemble a hidden forest of knowledge in need of protection, a tangled jungle of history and identity that lies, like the riverbones lie, largely beneath the surface of Saramaka.

It is late afternoon when I arrive back at the Gaama's office. Samson is here, speaking rapid-fire with Basia Joe.

"Any news?"

Samson reaches for the *sender*. He speaks a series of call numbers into the hand-held microphone. "I try all day. I call Brownsweg, but no one answers." He says the numbers again and gives one of the dials a twist. The spit and crackle of empty air. "I call city now." More numbers. Finally, a voice echoes back.

Samson speaks with Paramaribo and I sit down beside Basia Joe.
"You are a good cook," he says.
"Thank you."
"You should live here."
"Yeah?"
"You could be my wife." At this, Basia Joe heaves with laughter
and slaps me on the back.

Samson bangs the microphone back onto its perch. No one in
the city has heard from the Gaama. Everyone assumes he's still at
Rosebel but nobody knows for sure.

After dinner, I sit with the Aboikonis in their kitchen. I give Melissa
one of my empty notebooks and help her translate words from
Sranantongo into English. She will need English if she wants to
work with tourists. Mione sits in the doorway and sings *seketi* songs,
lyrical ballads of love and catastrophe, hope and despair, while
Sameni cooks up yet another batch of fried cassava.

I have decided to leave Asidonhopo. I still feel welcome here,
but the signs – they're telling me to go. No one knows where the
Gaama is or when he's coming back. My food has run out, I have
little money left, and my body is breaking down – the fungus has
spread to my arms. Jules has forgiven me, but I still feel horrible for
shaking him so hard. I'm not sure how much more of this endless
waiting and haunting slowness I can handle.

I came here to speak with the Gaama, to ask him a question
about his people. But over the last week I've learned the foolishness
of this, that whatever I am waiting for here, be it understanding or
closure, must be earned and will not be freely given. My education
in Suriname is only just beginning.

If the Gaama had been here to answer my question, to list off
the things his people need and profoundly deserve – legal title to

ancestral lands, for example, or full recognition of Saramaka personhood and nationhood under Surinamese law, or meaningful reparations for the drowning of half their homeland and countless other human rights abuses over the last fifty years, or a way to heal the cultural amnesia that, slowly but surely, is sweeping up the banks of the Suriname River – what would I have done with it? I would have written it down, but what then?

Social justice for Saramaka and all Surinamese Maroons will not result from a simple list of grievances in my pocket. Social justice will only be achieved when the Maroons are sufficiently in control of their lives to fight, once again, for change and freedom – just as their First-Time ancestors did, just as the Jungle Commando did. First-Time ideology preaches that "First-Time kills," and that "those times shall come again." Well, those times are here – in truth, they never ended – but with one crucial caveat: the time for violence has passed.

This battle is already underway. Recently, a long-awaited ruling from the Inter-American Court for Human Rights found the Surinamese government guilty of violating several articles of the American Convention on Human Rights in its dealings with the Saramaka community, specifically those pertaining to property rights, juridical personality, and judicial protection. The Court made a number of unanimous demands, the foremost being that legal title to ancestral lands, including absolute ownership of the natural resources both above and beneath the ground, be granted to the Saramaka people. Considering the profound effect this ruling would have on the Surinamese status quo, though, it is unlikely it will ever be implemented.

When I tell her I'm leaving, Sameni yelps and drops the lid of her pot. At first I think I've offended her, or that she will miss my presence in her house, or perhaps my cooking. But then I realize Sameni is simply disappointed her husband hasn't returned in time for us to speak. I'd like to explain that it's OK, that she and her family have taught me a great deal, but again, I don't have the words.

Mione tells me there is a boat coming tomorrow from Dangogo. He's sure they'll take me downriver. Then he grabs a can of pesticide from the windowsill and starts drumming on the cap with his fingers. The beat is remarkably fast and accented by a distinctly Saramaka rhythm. He sings a new song now, a song about his father, and when he reaches the chorus we all burst out laughing:

No one knows where the Gaama went,
I hear they lost the Gaama.
No one knows where the Gaama went,
They must have lost the Gaama.

In the jungle, a leave-taking is always an anticlimax. Journeys begin early, at cock's crow, before the river has shed the night. Villagers have barely woken and no one is himself. Good wishes lack heart, handshakes are flimsy, waves goodbye seem accidental – our paths have not yet crossed today and no one, it seems, will be missed.

The river breathes a cool mist and I can feel the fog in my lungs. Sameni, Mione, Melissa, and Jules wave from the shoreline as my boat pulls out to the current. None of their bodies look real but I wave back anyway, hoping they remember my visit. I must look like a ghost to them, slipping away in the grey.

The day begins at the fork in the river. Our boat fills with Saramaka bound for Paramaribo. Two particularly aggressive young men climb into the bow and yell at me. "*Foto bakaa! Foto bakaa!*" They recognize me, the phantom white man from the city.

The ride north is smooth but my body is tired. The fungus has spread from my arms to my armpits and down my legs, the chiggers on my ankles have begun to weep, my hands are covered in pus-filled

blisters, and my hamstrings ache. I slip in and out of uncomfortable dreams. A white, heron-like bird dips its beak into the shallows
and impales a fish. A pair of Amazon kingfishers follow us for more
than an hour, flitting ahead then falling behind then flitting ahead
again. Three boats overtake us, the men inside roaring and shaking
their fists as they go, all of them dressed in city clothes. They look
like tourists, like forgettable foreigners, but they are not tourists.
They are Saramaka who've grown tired of waiting, headed to the
city to make money any way they can.

19

THE HEALING OF ABENA-FOLO

We reach Abenaston as the sun is setting. I find *Abena-folo* and *Badja-mai* at their kitchen table, chatting with an old Saramaka man named Pa Jay.

"Tell us you've got more cookies," says Rachel.

I dig the last package from the bottom of my pack and hand it over. Rachel tears it open and gives a purple, cream-filled cookie to Ariel, who is smiling like a child.

"Watch this," says Rachel.

"Won't that put her in a coma?" I ask.

"Nope. *Abena-folo* is cured."

Ariel bites into the cookie and gives an exultant sigh. Pa Jay giggles and nods his head. Then Ariel reaches for her Nalgene. The bottle is filled with a murky concoction. Through the dark plastic I glimpse a feather, a few elongate leaves, chunks of black bark, and a handful of palm nuts suspended in a thick liquid. The nuts are the same species my monkeys used to eat. The sludge looks as if it's been scooped from a swamp.

Ariel takes a short sip and grimaces at the taste. Then she smiles up at me and pops the rest of the cookie into her mouth.

"This is why I love Suriname," she says. "My first cookie in five years and it's grape-flavoured."

While I was upriver, Pa Jay took the girls and their friend Dave across the river to Gengeston. Apparently, the people of Abenaston had grown tired of Ariel not being able to eat bread or drink beer with them, so Pa Jay suggested she visit the Gengeston *bonuma*, or medicine man, to see if he could help her with her gluten allergy.

Ariel had expected a "gnarled old man wearing a traditional *kamisa*," so she was initially suspicious when she was introduced to Djompei, the *bonuma*, who wore an orange polo shirt, khaki cargo shorts, and a pair of mirrored sunglasses. After a formal Saramaka introduction, during which the two men traded obligatory abstract stories and evasive metaphors, Pa Jay explained Ariel's affliction. A mild argument ensued. The men discussed the virtues of Christianity versus the old ways of *winti*. Finally, they settled on the clever compromise that God must have put the plants in the forest for his children to use, so it was OK to continue.

Djompei warned Ariel that he had already "pulled" many *tjinas* – a word that translates, according to the context, as either an allergy, aversion, sickness, inherited spiritual disorder, or demonic possession – but he had never tried to heal an allergy specific to flour. Perhaps to demonstrate the challenges involved, the conversation turned for a moment to the man in Pamboko who is working on a *busi dresi*, a medicine, to make his wife ugly to the eyes of other men. His progress has been slow, apparently, because he is having difficulty procuring the requisite hair from a jaguar's cheek.

Djompei's daughter, Sita, brought a calabash filled with leaves and water. On top sat a small shell stuck through with a parrot feather. After speaking a few Christian and Saramaka prayers, the witnesses formed a tight circle around Ariel. Djompei took a small sip from the calabash, followed by Pa Jay, and the *bonumu* passed the feather over Ariel's outstretched tongue three times. Then he brought the calabash to Ariel's lips and poured the liquid down her throat.

"It was sort of fizzy," she recalls. "Granular and chunky. And then came the scary part. They had sent Dave on a scavenger hunt for some bread, and now they handed me a chunk of this godawful packaged Twinkie-like sponge cake thing. My first real bread product in several years. They had me take a nibble of the cake, then a sip of the brew, then a nibble of the cake, then more leaf water. Back and forth few times. Then Pa Jay and Djompei spoke a call-and-answer ritual to protect us from any ill effects of the medicine, a sort of spiritual payment for the gifts of the jungle."

After the ceremony, Djompei instructed Ariel to keep the calabash topped up with water and to drink from it constantly over the next eleven days. He also told her to gradually increase her bread intake. She barely made it back to Abenaston that night before she became terribly ill. But the next morning she felt great. She has been eating bread and drinking beer ever since.

The next morning, the girls accompany me to Atchoni. We drink a celebratory Parbo beneath the same shelter where I met Wout a week earlier. We toast the old pastor, the man who believes Maroon medicine men are devil worshippers. Ariel obediently chases each sip with a short swig from her *busi dresi*. We tell the man at the chicken stand about the allergy-pull and he smiles proudly.

"Saramaka *sabi*," he says. The next round is on the house.

The return trip on the Atchoni Road is cruel and exhausting. I sit with thirty Saramaka and a few tourists in the back of an old cattle truck fitted with wooden benches attached to a steel scaffold. Every space not filled by a human body or gasoline canister is taken by a rainforest creature: a jungle turtle, still alive, with a stick thrust through its arm holes; two red and blue macaws who, thankfully, are quietly sleeping; a huge *anyumara*, long dead, the fish wrapped in clear plastic. A few of the windows have been smashed and are now draped with ratty tarps; when the rains come we are all soaked by bauxite sludge. Newborn babies dangle from their mothers' laps. Mothers slip in and out of slumber. A little girl in a yellow dress sits across from me, hugging a plush stuffed animal.

On the outskirts of Paramaribo, young boys perform backflips in coconut groves and young girls turn cartwheels in ditches. Overgrown fields are strewn with white plastic furniture and the dusk-lit air roils with trash-fire smoke. The truck lets us out on Saramaccastraat and I stumble across town in a daze. I pass a homeless man prostrate on his cobblestone bed, a candle flickering in the window above him. In the shower at Stadszending, half-asleep and travel-numb, I wash the Red Road from my skin.

The next morning, at the internet café, I find an email from Emma. *I hope all is well and you have found what you are looking for.* She warns me to think long and hard about where I will live when I get back to Canada. *Toronto or Vancouver?*

Then I hear the startling news that broke while I was upriver. Twenty-three years after the fact, a preliminary tribunal has ordered Desi Bouterse and his military associates to stand trial for the December Murders.

Bodi meets me at Stadszending just after dusk.

"Big Man not happy, *toch*," he says, handing me the CD.

"What do you mean?"

"Too many combinations." Bodi stares up at the lowering sky, his jaw rippling.

"We gave him what he wanted," I say.

"No. We give too many."

"Too many?"

"He wanted a thousand. This has more than 350 thousand."

"But that's how many there are!"

"The number one to twenty-six, *toch*."

"Yes!"

"Too many."

The cleaning lady, a cross-eyed Javanese woman, arrives with fresh towels. Bodi tells her to keep walking.

"So he didn't pay you?"

"No."

"That's shit, Bodi. We gave him what he asked for."

"He says we didn't."

"He probably copied the disk before he gave it back to you! We're getting ripped off here!"

Bodi sneers. "Maybe you can talk to him."

I light a Morello. Bodi lights a joint. The security guard down at the front gates is fast asleep.

"Army fire me," Bodi says.

"What? Why?"

"*Mi no sabi.* Clean up? *Mi no sabi.*"

I look at Bodi's hip. The light on his mobile is out.

"Shit, man. I'm sorry."

"*Alamala broko,*" he says. Everything is broken.

"What will you do for money?"

"*Mi no sabi, toch.*"

Suddenly Bodi doesn't look so tough anymore. My impression of him as a disciplined bodyguard and shadowy spy is replaced by something entirely less impressive. With his yellow polo shirt, mesh ball-cap, Adidas sandals, and slight pot-belly, Bodi looks more like a soccer dad than a plainclothes commando. Maybe I've been kidding myself this whole time. Maybe those scars on his hands *are* the remnants of some disease.

"Did you hear about the tribunal?" I ask.

"Ai."

"Maybe now you'll get the truth."

Bodi smiles. "Everyone knows the truth."

"They do?"

"He did it. Maybe he pull trigger, maybe he doesn't. But Bouterse kill those people."

"So he should pay," I say.

Bodi laughs. "Before revolution, Suriname has kind of apartheid. Someone should pay for that, too."

"You will have closure. You can move on."

"Fucking whites," says Bodi. "Always tell people what we need. Bouterse is old, Suriname is young. How it help to put him in jail?"

"It's not about Suriname. It's about the victims, their families."

"It's about Suriname, *toch*. And politics. And the Dutch. And money. World wants us to punish, to forgive. Suriname wants to forget."

"You mean *you* want to forget."

Bodi stands up, stubs his joint out. "I *am* Suriname." He walks along the terrace to the stairs. "I see you later. We play football."

"Maybe not," I say. "I'm going home."

He walks back, reaches out to shake my hand.

"Where you live in Canada?"

"I'm not sure yet," I say, trying to smile. "Maybe Toronto. Maybe Vancouver."

"Maybe I come visit."

"That would be good."

Bodi walks to the stairs. Halfway down he stops. "Or maybe you should stay here," he says quickly, surprised by the thought. "Maybe Stadszending is your home now."

Joseph of Nazarene finds me on the Waterkant two hours later. I am three *jugos* in and very drunk.

"Never trust a woman," he says, pulling up a chair. "Woman monopolize your finance."

Joseph and I first met four months ago in the Palmentuin, the palm garden where he lives behind the Presidential Palace. I had gone there in search of a troop of injured monkeys rumoured to pass through every morning and every night. "I came to Suriname to open the hearts of men," he'd told me back then. "My Babylon name is Brian. My godly name is Joseph. I live here because I have a deadly enemy."

Joseph carries himself as a Rastafarian, with shoulder-length dreads and the ratty clothing of the homeless. He swears, however, that he is not Rasta. His deadly enemy is his old boss at the waffle house where he used to work, a man he calls simply "Pancake." He also claims the gift of prophecy and the ability to perform miracles.

"Name me the miracle of Christ," he says, helping himself to some beer.

"Water to wine," I say.

"That's not a miracle. That's easy – just add fruit. Name me a real miracle."

I can't think of one.

"These people," he says, gesturing at the drinkers around us, "they are not of God. Why couldn't the Romans recognize Christ?"

"Because Christ looked like everyone else."

"Precise!" Joseph reaches across and grabs my hand. The word "SHAKE" is tattooed with cashew oil on his forearm.

We chat for a long time. He tells me how he used to run guns for Brunswijk, how easy it was for a man with wisdom to cross over from French Guiana with a satchel of weapons on his back. Then he tells me how he spent his money.

"Brazilians, Dominicans, Javanese, Hindustani, Creole. One hundred SRD an hour. Two hundred the night. Beautiful women." Joseph upends the last of his beer on the table, rubs his fingers through it as if reading tea leaves.

"Where?" I ask.

"Just down street."

"Where?"

"Club Aventura. Near Pro-Ice."

We drink without speaking. Joseph stares at the spilled beer, his forehead creased in thought. Then he slaps the puddle with his palm. "Lovers come to the garden and I talk with them," he says quickly. "I teach them the secrets. I teach them forever."

"Do you ever see the monkeys?"

Joseph smiles. His teeth are ridges of small yellow hills. "Yes. I talk with one last week. I have three Cayenne bananas, and one of the monkeys wants one. He comes down so close I can see the gash on his face. He is half-fur, half-skin, yes? So I say, 'I will give you banana if you come and take it from me.' And Jah alive, you know what he did?"

"He jumped down and took it straight from your hand."

"Precise! That monkey walked on two feet!"

Club Aventura is half-filled with eager men. The girls sit bored at the bar, smoking and chatting with these men, the ones they are friends with, the ones who come early to watch the fights. On the television, massive warriors are getting their faces pummelled, their arms pulled out of their sockets, their legs twisted into submission, their senses knocked free from their skulls. The bar slowly fills with businessmen and Chinese teenagers. All of their eyelids are swollen like mine. A few of the youngsters disappear into the back and return a half-hour later, their chests pumped, their fists clenched, but their eyes still young and wide. I once read that sex is the most popular pastime in Suriname.

A tall Brazilian named Adriana sits down next to me. She has long black hair and a silver stud in the middle of her chin. She writhes like a snake in her seat. She asks what I want and I tell her booze. She laughs and thinks I don't understand.

The night speeds up, the warriors fight on, and the room fills with cigar smoke. One of the girls, a gorgeous Colombian with a gap in her teeth, performs a striptease on the table in front of me.

My vision swims. A blurry man sits next to me and hands me a Montecristo.

"Ask me how many people I have killed."

"Alright."

"Twenty-seven. But I don't do that any more."

"Why's that?"

"They don't ask me."

The dance ends with a thick Creole, a military officer, pretending to bury his face in the girl's crotch. When she climbs down from the table, the officer pinches her ass, and suddenly every girl in the room is screaming. The men laugh and so does the stripper, but the rest of the girls are livid. "Don't set that kind of example!" they yell in Portuguese. "Get your hands off our sister!" I sit with my head in my hands, the screams and laughter booming in my ears. I order double whiskeys and the warmth makes me think of snow, sleet, slush, Toronto in winter.

Adriana reappears. She asks how much money I have. I stand shakily and let her search my pockets. She spreads my coins and bills on the beer-soaked table. Ninety-seven SRD. I nearly collapse but she catches me.

"So close," she says.

She gathers my money, takes my hand, and drags me into the bathroom. Then she turns my empty pockets inside out.

"So close," she says again.

I say something back, something slurred that makes me laugh.

"Huh?" she says. "What does that mean, 'in imaginary country'?"

We stumble into the back room and up a flight of stairs.

We're in a child's room, a double bed, two bunk beds, and cartoon wallpaper. Before she steps into the shower, Adriana tells me she lives here with four of the girls downstairs. I collapse on the bed and shut my eyes. My mind floods with a river of libations. I've heard about the single-engine planes filled with Latino girls. I know about the horrors, the crushing debts, the modern-day slavery. But this is not that. The love in me is twisted, but this is not that. I am a reservoir of good-hearted gifts. I am not really doing this. In an imaginary country it is impossible to act like yourself.

Adriana steps from the shower, a towel wrapped around her head. She is naked and gorgeous and smells like bird-of-paradise.

I am nearly sober when I leave the club. The streets are empty, flooded with rain, and my stomach aches with booze. There is a message on my phone when I get back to Stadszending. It's Bruce Hoffman. In three days he is flying south to the Trio Indian village of Kwamalasamutu, near the border with Brazil, to continue his studies with the shamans there. Kwamala is where Fritz von Troon promised to deliver me. Bruce will clear it with the paramount chief. He has offered to take me with him.

PART THREE

The People Here

"We should insist on a recognition of the mystery, the miracle, and the *dignity* of things, from frogs to forests, simply because they *are*."

CURTIS WHITE, *THE IDOLS OF ENVIRONMENTALISM*

"When you are lost to ones you love, you will face south-southwest like the caged bird."

ANNE MICHAELS, *FUGITIVE PIECES*

20

—

THE WISDOM OF BONES

I remember a story one of the timber framers at Raleighvallen told me. A few years ago, Gene had been working as a medic on a wildfire in Zuni Nation territory, in western New Mexico. Just when the blaze seemed to be under control, the last stubborn flames licked up a steep slope and leaped the road, where the brush was still thick and dry. The injuries started pouring in – broken bones, sprained ankles, third-degree burns. During the worst of it, a young Zuni boy emerged from the smoke with the worst thumb dislocation Gene had ever seen – "I mean, bent right into his palm." Gene told the boy to wait because many others were worse off.

Then a Zuni elder appeared at Gene's side. He asked if he could speak to the boy with the injured thumb. Gene pointed him out and the man disappeared into the fog. Forty-five seconds later, the boy returned and showed Gene his hand. "His thumb was perfect," said Gene. "No bruising. No swelling. Nothing."

Once the fire was put down, Gene searched the smoke for the Zuni man. He eventually found him down the road and asked how he'd healed the boy.

"When I was six years old," said the man, "I was struck by lightning. Since then, I have had the wisdom of bones."

He appears pushing a wheelbarrow filled with flesh and blood. In the air to the north, the Russian-built Antonov howls its retreat, 350 kilometres of shadowy jungle between it and Paramaribo. I've been in Kwamalasamutu ten minutes and already the shaman has found me. His name is Amasina. He has the flattened features and glossy black hair of a pure Amazonian. He wears the ripped T-shirt and torn jeans of a hard and busy man. Bruce introduces us and Amasina asks if we want to buy some meat. I have never eaten tapir so I buy a chunk. Gunshots ring out somewhere in the village. The meat is purple and warm against my palm.

Bruce and Amasina talk for a while in Trio. They have been friends for a long time, two men whose cultures couldn't be further apart but whose interests couldn't be more closely aligned. Amasina is one of the senior shamans in Kwamala, an expert on the medicinal plants found in the forest here. Bruce is here to record what Amasina knows before Amasina dies. If the Baptists had never arrived here and laid waste to Trio culture, Bruce and Amasina might never have met.

Bruce switches to Sranantongo and tells Amasina I'd like to speak with him. Amasina nods. He drops his wheelbarrow in the shade of a nearby hut and I do the same with my bags. Then he motions for me to follow him as more gunshots boom.

Soon after his adventures with Fritz von Troon, Mark Plotkin returned to Suriname for a second field season. It was 1982, the time of the December Murders, and the country was teetering on the brink of civil war. Desperate to escape Paramaribo, Plotkin found a bush pilot willing to take him to the remote south-southwest of Suriname, far from the guns of war. The pilot eventually dropped him here, in the multi-ethnic village of Kwamalasamutu, where, he says, he "stumbled onto an ethnobotanical gold mine."

With the help of a friendly Waiwai teenager named Koita, who translated for him, Plotkin used Kwamala as his base. Kwamala is where Plotkin learned a recipe for curare; where the great Jaguar Shaman haunted his dreams and allowed the young scientist to study his secrets; where his rational view of the world was repeatedly challenged by the "black arts" and "spirit realms" of shamanism; where he learned how limited Western pharmacology is compared to the vast botanical wealth of the rainforest, and the vast knowledge of the medicine men. "The typical Amazonian shaman," he writes, "served not only as a physician but also as a priest, pharmacist, psychiatrist, and even psychopomp – one who conducts souls to the afterworld." These men are master healers, he says, medicinal savants able to soothe all physical, psychological, and spiritual wounds.

I follow Amasina down a short hill into the bush. Only now do I realize how small he is, his body economical, his skin tight and smooth. The shaman stops on the path to show me his plants. With hand gestures and a few words, he explains how each one can cure me. The one with spines like eagle talons will rid me of the common cold. Others will cure the pain in my back and the white spots on my face. The last one will help induce childbirth. For a moment, I wonder if Bruce has misled Amasina into thinking I'm also a scientist. Ethnobotany is not exactly why I'm here.

At the top of a small rise we emerge into a clearing between four thatched-roof huts, where a makeshift shelter has been erected. About fifty Trio men, women, and children sit beneath it on benches, watching Surinamese reggae videos on a black and white television. In the middle sit two huge barrels of *casiri*, beer made from cassava bread that has been chewed, spat into a vat, and left to ferment for days. From time to time, a woman arrives with a bucket of fresh brew, which she empties into the barrel. A man sits with his rifle next to him, the source of the gun blasts. This is a birthday party for a twelve-year-old girl.

We sit in silence and watch the party. The beer is ladled out and brought to each of us in turn. Beside me, a mute boy makes erotic hand gestures at his brother, whose eyes are glazed over with booze. Every now and then, a teenaged girl hurls a handful of gumballs and Hall's mints onto the ground and the children riot. Soon it is my turn to drink. One of the girls brings me the ladle. Following Amasina's lead, I swallow it in three gulps. The booze is pink and warm and sour.

I don't know how to broach the subject with Amasina; I'm afraid he'll get angry, or he simply won't understand. I decide to leave out the details. I don't tell him about my travels, my sadness, my shameful attempts to heal. All I say is I need his help to purge a woman from my mind.

Whether Amasina understands is unclear. What is clear is that he cannot help me. "*Mi no luku Gado moro,*" he says. He does not visit God any more. The Baptists consider that to be devil worship, punishable by whippings, so the shamans here have stopped visiting the spirit realms.

Amasina shakes his head and frowns. He might have helped me twenty-five years ago. I've come to Kwamala a quarter-century too late.

Disappointed, drunk, I stumble through the village. At the north end I find the Traditional Medicine Clinic, the original home of the Shaman's Apprentice Program.

Early in his research, Plotkin realized Amazonian shamanism was slowly dying. Fewer and fewer young people were dedicating their lives to the arcane ways of the medicine man, and for Plotkin this signalled a coming tragedy. "When a shaman dies in a preliterate culture," he once said, "it's like a library burning down."

Plotkin realized this was a tragedy not only for the Amerindian communities themselves but for humanity as a whole, since the preservation of Amazonian botanical wisdom is a critical step to preserving
the Amazonian jungle itself.

Determined to help the Trio pass on their knowledge, Plotkin
founded the Shaman's Apprentice Program, a scholarship course
in which young men and women learn about medicinal plants and
their recipes from the senior shamans in their village. By reviving
shamanic knowledge in these communities, the program demon-
strates the profound relationship between humankind and the
botanical world. The program now operates in numerous villages
throughout the tropics. This clinic in Kwamala is where it all began.

Inside, I find a young Trio girl behind a desk, sorting through a
binder. She offers me a tour. In one of the back rooms she lifts the
lid of a refrigerator, retrieves a small bottle, and pours me a cup of
thin brown liquid. As I drink the pungent juice, she mutters the traditional name of the brew, which in my daze I fail to write down.
This one is good for the stomach, she says. She says nothing about
casiri overdoses.

I find a topographical map of Suriname tacked to the wall. I
place my finger on Kwamala and absently trace the route from here
to the savannahs, the home of okopipi, the blue frog I'm not permitted to see. I am feeling foolish that I've come all this way for
nothing; I'm feeling foolish that I'm still in this country at all. My
finger follows the Sipaliwini River southeast until it reaches a place
called Mamia.

Then an Indian appears at my side. He seems vaguely familiar.
"San yu luku?" he asks.
"Mi luku okopipi," I say.
"Yu wani si?"
"Ai." I say, smiling. "But STINASU wouldn't give me a permit."

323

The man touches my shoulder and grins. "I help you," he says. "My name Koita. I help you find *okopipi*."

I stare at this man through my *casiri* stupor, try to figure out why I recognize him, how he read my mind and if he's being serious. Then it dawns on me. This is Koita, the same man who helped Mark Plotkin twenty-five years ago. I recognize his face from a photograph in *Tales*.

"He was small and wiry and stood about five-foot-two," Plotkin writes about first meeting Koita. "Like the other Indians who watched me so intently, he had painted his bronze skin dark blue with the fruit of the *meh-nu* tree. Suspended from a white cotton string around his neck was a charm, a slightly curved, yellowish white jaguar tooth, about three inches long."

A quarter century later, Koita wears a pair of jeans, a purple T-shirt, an orange sun hat, no paint on his skin, and no charms around his neck. His face, though, has barely changed. On his forearm is a tattoo of his name, carved in the Trio language.

"You must ask Granman for information, not STINASU," he says.

"Information?"

"Yes. Only Granman can give you information to see *okopipi*."

"Information?"

"Yes!" he says, realizing I'm drunk. "I will set up meeting. Tomorrow you will meet the Granman."

At the north end of the soccer field, I find the Okopipi Pakoro guesthouse, or Home of the Blue Frog. Inside, representatives from Conservation International and an ecotour operator from Paramaribo have convened a meeting with the leaders of Stichting Maio, the village development committee. They are discussing plans to build a tourist lodge at Ewana Samu, a gorgeous white sand beach with natural hot springs a short boat ride from Kwamala.

Cyrano, the tour operator, is very excited about the project. "I just got back from Holland," he says. "I spoke about Kwamala for five days to investors. They are very keen."

"But we have heard this before," says a Trio named Kupias. "We have had meetings like this for ten years."

"But now we are serious," says Annette from CI-Suriname. "The money is there. The people are there. Now we need to talk about what the Trio want from the project. *San yu wani?*"

Another Trio, Thomas, takes the floor. "First thing, food and bread. Second, we want to sell souvenirs. The main thing? Money."

"But who will get the money? How will it be distributed?" asks Annette.

The Trio look at each other. "We need to decide," says Kupias.

"Before money comes, we have to agree on how it is spent," says Annette.

"Capacity equals knowledge plus skills," says the consultant from CI.

"We will use all natural building materials," says Cyrano.

"People think ecotourism will solve all their problems," says the consultant. "But it has to be a viable business first."

"We will build with wood, stone, and clay," says Cyrano.

"You have to be so careful with this," says Annette.

"Without future plans, without marketing, the lodge will sit and rot," says the consultant.

"Why hasn't anyone said anything about nature, culture, traditions?" asks Annette.

"Nature, culture, traditions," says Cyrano. "These are the basics of the project."

"Instead of shooting monkeys," says the consultant, "take tourists to see them."

"The solar project here, the water project here, both failed because of maintenance problems," says Annette.

"Maintenance, sustainability," says Cyrano. "These are the basics of the project."

"The government Tourism Ministry doesn't have a budget," says the consultant.

"There are only four hundred hotel rooms in Paramaribo," says Annette.

"We need open skies," says Cyrano. "We need another airline in Suriname."

"Kwamala is the future of tourism," says the consultant.

The Trio listen. Thomas tries to take notes but can't keep up. Finally, the others stop talking and the house goes silent.

"When do you want to see the sanctuary?" asks Kupias.

"Tomorrow morning," says Annette.

"OK," he says. "We go to Werehpai tomorrow."

The Trio leave the guesthouse and Cyrano sits down next to me.

"What have you found in Suriname?" he asks with a smile.

"Found?"

"Yes. You are a writer. What have you found?"

"I'm not sure exactly."

"Don't worry. Everyone says Suriname is a mystery to outsiders, but it is a mystery to people who live here, too. Suriname should be one big chaos, one big confusion. But in the end it is very much the opposite." Cyrano leans back and closes his eyes. "I am really very touched by this place."

Cyrano owes his life to the shamans of Kwamala. A few years ago he came down with leishmaniasis, the flesh-eating disease, on his shin. Western medicine could do nothing for him, and after nine months his bone was showing through the weeping wound in his leg. Then he heard about the clinic here in Kwamala. He sent a friend to ask the shamans for help. For three days, one of the elders brewed various mixtures of plant extracts, experimenting with the recipe. Cyrano's friend finally returned to Paramaribo and gave him the resulting ointment. In one week his wound was healed.

"My son is one-sixteenth Amerindian," says Cyrano. "But whenever he's asked who he is, he always says he's an Indian. It is something you cannot grab, Surinamese culture. But even though you can't grab it, you can still hold onto it somehow. Come with us to the sanctuary tomorrow."

"If there's room, I'd love to."

"I remember one day when I was young, I asked my father to walk with me in the bush. I wanted to show him how knowledgeable I was, how strong I was in the jungle. He took one look at my heavy boots and my sharpened machete and laughed at me. He said, 'If you walk this way in the forest you are going to be killed.' I was stunned. 'By who?' I asked. 'Who is going to kill me?' And my father, I will never forget this, he laughed right in my face. 'The forest will kill you,' he said. 'Take off your boots. Drop your machete. Walk where the forest wants you to walk.'"

Kwamala wakes at six o'clock to a disembodied voice. From a speaker perched high atop his office hut, one of the Granman's aides announces the day's events in a haunting monotone, the news wafting through the sleeping village like smoke. I feel as if I've woken in a deleted scene from *Apocalypse Now*.

Three hours later, the Werehpai expedition assembles at the shoreline in two broad dugout canoes. A half-hour up the Sipaliwini, we leave the river behind and hike into the jungle. The forest is wet, and we trudge through the mud without speaking. Screaming piha birds shriek their two-syllable greetings and pairs of caracaras warn us away from their nests. The electric buzz of cicadas rises on the late afternoon air, as does the sour stink of wild boar. Soon the trail is blocked by a series of massive boulders. We have arrived at Werehpai.

Our guide is Kamainja, a Trio who studied under the Jaguar Shaman and is now the proud protagonist of *The Shaman's Apprentice*, a children's book authored by Mark Plotkin. Before taking us any farther, Kamainja tells us how he found this place in May 2001. His tale has become the latest chapter in the canon of Trio mythology.

"I was chasing a herd of wild boar when my hunting dog disappeared," Kamainja says. "I looked for him all day but couldn't find him, so I returned to camp. My wife was very upset. Our dog is a good dog. I went to sleep angry and very sad."

In his dream that night, a voice described a secret pathway through the forest and told Kamainja to follow it. "You will not find your dog," said the voice. "Instead, you will find pictures of your ancestors carved in stone."

The next morning, Kamainja woke before the sun and returned to the bush. "I followed the same route I'd learned in the dream. I secretly expected to find my dog, but by noon I had found nothing so I sat down to eat my lunch. I was lost. I did not know where I was. Then, on the surface of a stone at my feet, I saw a face staring up at me."

Kamainja looked up and realized he was sitting at the base of a towering complex of granite boulders overgrown with moss and vines. He had never seen this place before. "The face told me to explore the stones," he says. "I did what the face told me."

Among the boulders, Kamainja found a labyrinth of seven rooms, each with crude petroglyphs, haunting images of faces and rainforest animals, carved into the rock walls. The sloped floors were carpeted with shards of pottery; the sound of them splintering beneath his feet frightened him. In one room there was an altar stone, its flat surface thick with the bones of wild boar, armadillo, and monkey. Once home to his ancestors, this room was now a jaguar's lair.

Kamainja left quickly. He paddled back to Kwamala and alerted the Trio Granman to what he'd found. Government officials and local archaeologists were told and the site was declared a sanctuary. The archaeologist Abelardo Sandoval arrived from the Smithsonian to conduct a preliminary dig. Representatives from Conservation International began exploring the ecotourism potential of the site. Fewer than one hundred people alive have set foot in the houses of Werehpai. Until now, only two people have been permitted to photograph the site.

After a short prayer, Kamainja leads us into the labyrinth. We enter by the light of our headlamps. The air is dank and cold, almost subterranean. The floors are slippery with tropical lichens and dotted with fragments of clay.

In the first room, there are carvings of stick figure bodies with oval heads, their mouths wide open; of giant spiders and coiled serpents, threatening to strike; of half-man and half-butterfly creatures, loosing arrows; and scattered throughout, images of men with ornate headdresses, suggesting Amerindian royalty, and women with simple pigtails.

The carvings come alive in the skittering light from our lamps and cameras. We keep walking, and in every room, I feel that a hundred eyes are on me at all times, following me as I move through the shadows. No one, save a jaguar or two, has lived here for at least a millennium, but the caves still feel eerily inhabited.

Werehpai is at least five thousand years old, a prehistoric apartment complex of naturally occurring stone shelters that most likely served as a temporary village for nomadic hunters and small tribes of agricultural Amerindians. Kamainja's discovery is easily the

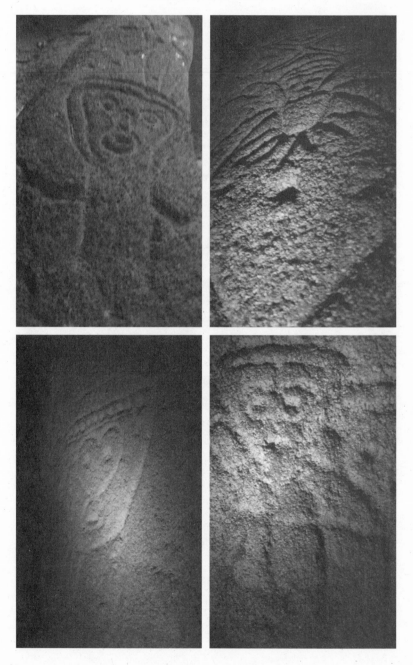

most significant archeological find in Surinamese history, and the petroglyphs are one of the most remarkable examples of stone-age artwork in South America.

The passageways between the rooms are narrow and dark, claustrophobic, and we often have to crawl. Kamainja rushes us through the maze, clearly uncomfortable with the pops of our flashes. "*Snel, snel,*" he says, urging us to hurry. He is very proud of this place and intensely protective of it.

One room is off limits under Sandoval's instructions. This is where Kamainja found the altar stone. Apparently, the petroglyphs there are the most impressive, the walls teeming with ancient stories told with hammer and chisel.

An hour later we emerge from the dark of the seventh house. As our eyes adjust to the muted light of the jungle, Kamainja explains the site's significance to the Trio and how it got its name.

Trio legend says that Werehpai was a young woman from the Akijo tribe, a race that lived for thousands of years in the remote jungles of the Upper Amazon Basin. The Akijo were an advanced people – they knew how to paint, to draw, and to weave – but they were a warring tribe, renowned for their violence and cannibalism.

One day, Werehpai was given two children to raise as her own, a brother and sister who had been kidnapped from a neighbouring tribe. When the children reached adolescence, the sister was removed from Werehpai's care and subjected to a horrific Akijo ritual – she was tattooed from head to toe and then kept alive as her limbs were shorn off and devoured by her tribespeople.

Terrified and heartbroken, Werehpai helped the girl's brother, Aturai, escape into the jungle. In time, Aturai returned with an army of men, bent on revenge. A vicious battle ensued. The jungle spirits, who could be trusted to side with the righteous, joined with Aturai. When the jungles fell silent, he had prevailed – the Akijo had been destroyed. The Trio believe the petroglyphs at Werehpai are an attempt by their ancestors to create a visual record of that battle – the graphic novel of a legend that has been passed down orally for thousands of years.

Soon after the discovery became public, the Trio held a feast to bless the site. They offered tributes to their ancestors, and to Kamainja and the dream that led him here. His tale of courage triumphing over fear has quickly become legend here and now affords him a measure of celebrity status among his people. It proves that even today, in parts of the world where good storytelling is still paramount, new myths are continually being dreamed, lived, and spoken.

We hike back to the Sipaliwini along the tangled path of Kamainja's dream. As we climb into the boat, I ask him if he ever found his dog.

"No," he says with a smile. "The dog found me." A week after he stumbled upon Werehpai, Kamainja's hunting dog appeared at his door.

Over the next three days, Koita and I have three meetings with the paramount chief of the Trio Nation, Granman Asongo. At the first, Asongo asks me why I want to see *okopipi*. I tell him about my travels, my fascination with his country, my growing sense that *okopipi* might be what I've been searching for this whole time. I also mention the conservation officials who denied me the permits. Asongo smiles. This is Trio land, he says. Here in the south, the Granman has the last word.

At our second meeting, Asongo asks again about my intentions, and I repeat my script from the previous day. Then he tells me I must write a book about Kwamala. To bring tourists, he says. To bring money to the Trio. Come back tomorrow, he says. He will give me his decision then.

I am feeling confident, so I spend the rest of the day renting a boat and a motor and spending every penny I have on gasoline. I pack my things and hire a crew. Their names are Mawa, Lukas, and Ipiroke, young Trio men who haven't worked in months and who speak slow, broken English. I have given up on finding a way out of sadness and am now determined to find this frog.

The next day, our final meeting is short. Before I've sat down, Koita turns to me and says, "The Granman gives you information." I stare at him for a second, completely confused, and then realize his translation is a bit off.

"You mean permission?" I ask.

Koita covers his face and says something to Asongo, who laughs.

"Yes! Permission!" says Koita. "The Granman gives you official permission to travel into the savannahs and visit *okopipi*." Accompanying us will be the Basia, an immense, goofy-looking man who will represent the Trio government on the expedition and report back to the Granman.

Asongo sets two conditions. First, I am not to take *okopipi* home with me to Canada. Second, I am to bring the Granman as many red-footed savannah turtles as I can find.

"How long is the trip?" I ask Koita.

"Sunset, sunset, sunset, arrive."

THE BLUE JEWEL OF THE JUNGLE

I wake at six o'clock to the soft murmurs of the Trio language. Outside, my crew has gathered under the overhang of my thatched-roof hut. Lukas and Mawa are inspecting the motor, which stands balanced on its prop, the engine casing in pieces on the ground. Mawa holds a small bottle of oil and Lukas mumbles to himself as he fiddles with a wrench. Ipiroke and the Basia lean against two enormous gasoline barrels and frown at the ground. The sun has risen. On the horizon, the airstrip is beginning to glow.

"*Kuday mana*," I say, the Trio morning greeting, the only words I know.

The men look up all at once. "*Andu!*" yells Ipiroke. "*Fa yu sribi?*" asks Lukas. Mawa drops his oil bottle and walks toward me. The Basia just smiles his goofy smile.

"We have problem," says Mawa.

"I see that," I say.

"Man we pay for motor not happy. Now we use my motor. First we fix."

"How long?" I ask.

"Not long."

"We leave seven o'clock?"

"Yes. Seven o'clock."

At five to ten, Lukas and Mawa finally screw the engine casing back into place and we carry the equipment down to the river. We have a lot of equipment. Three waterproof barrels filled with cassava bread and flour, three hundred litres of gasoline in jerry cans, two old detergent buckets overflowing with *casiri*, a bright blue jug with one hundred litres of rainwater for drinking, fifty kilograms of rice, two tarps, two storm lanterns, six life jackets, two rifles, one bow and a quiver of arrows, an extra outboard motor. And all of our hammocks and clothes and boots in countless garbage bags. We will be gone for a week. Aside from the staples, we will eat whatever we can kill.

Our boat is a twenty-four-foot dugout, a single tree trunk hollowed out by fire and machete. Along the inside, metal scraps have been nailed over holes in the hull. Fully loaded, the boat sits low in the water, our mountain of equipment bulging over the gunwales.

We push off and the crowd onshore yells their encouragement. A young girl is rolling a huge round of cassava bread beside her like a hoop. The bread is as tall as she is. When it catches the sun I can see right through it, a golden-brown orb with the shadow of a girl in its centre. In a hammock nearby, the mute boy from the party three days ago sees me and waves. He makes a series of threatening hand gestures and points to the south, as if we are crazy to be headed that way. As if danger lurks upriver.

Then I spot the extra outboard motor sitting in the grass.

"Lukas," I say. "*Yu fergiti* motor."

"No problem," says Lukas, yanking on the ignition cord. "Motor good now." After the seventh pull, the engine roars to life.

Prior to the 1960s, the Trio lived in small, semi-nomadic villages of twenty to forty inhabitants scattered throughout southern Suriname and northern Brazil. Their territory was the watershed divide between the Sipaliwini Basin, where the rivers flow north to the Atlantic, and the Upper Amazon Basin, where the waters flow south to the world's largest river.

The Trio are made up of historically separate aboriginal groups. At some point, likely long before the New World was discovered, these peoples began referring to themselves inclusively with the broad denomination *Tareno*. Over time, outsiders evolved this word into Trio. *Tareno* means "the people here."

In 1954, the protestant West Indies Mission established the Suriname Interior Fellowship. Six years later, the Surinamese government began Operation Grasshopper, building forty new airstrips to open up the country's tangled interior. These two events marked the beginning of the end for the traditional Trio way of life.

Along the shore of the Sipaliwini the jungle is thick and seems to brood. Young cecropia trees lean out over the water like curious children, their crowns of convoluted leaves drenched in sun. A white, heron-like bird with a bright blue beak watches us pass and then plunges its head into the river.

Soon the river narrows and the rapids of Gruni Keni appear. With a few less inches of water we'd have to get out and carry the boat. Mawa stands in the bow and directs Lukas through the stones. We get stuck on the bottom a number of times and Mawa digs a long stick into the riverbed to punt the boat forward. We are barely moving. The motor chokes when Lukas cranks the gas.

A boat approaches from upriver carrying a Trio family, the husband at the motor, his wife and two small children in the bow. Their boat is as heavily laden as ours but is a better vessel, shaped more like an arrow. As they speed between the rocks, aided by the current, Lukas yells something to the man, who quickly turns his boat around and meets us at the shore. After a short conversation, we follow the family downriver.

The family has agreed to switch boats with us. The husband laughs at the thought of us attempting this journey with our dugout.

His family unloads their possessions onto the shore, a menagerie of dead or dying animals. A live turtle with a stick through its arm holes. A baby kinkajou with a rope around its neck. River fish strung through the gills with a piece of wire. Two blue and green parrots, their feet tied together with string. An adult spider monkey, his eyes closed, his hands limp, a single bullet hole through his chest.

Lukas lifts our faulty motor onto the stern of our new boat as we load our gear into the hull. Within minutes we are back on the water and the family has disappeared into the jungle. We make it up Gruni Keni despite our engine's complaints. Then the river calms. Fist-sized balls of cotton fall from the *kankan* trees and drift on the river's surface. We pass several orapendula colonies, the black birds flashing their yellow tails as they dive for cover into palm-thatch nests. I count twenty-three nests in the canopy of one tree, each one hanging like a black teardrop.

Two hours later, we round a bend in the river and hear the ominous sound of rushing water. Half a mile up ahead, the river is full of boulders and whitewater. This is Ewana Tepu, or Iguana Falls, the largest rapids on the Sipaliwini. There is no way our engine can handle this, I think, as the boat slows and Mawa stands to scout a route. Lukas steers us to the right-hand bank, where a narrow side channel empties into the river.

We enter the channel and are enveloped by a close, green darkness. Our route is less than five metres across, a shallow jungle stream that will surely dry up in a week or two. At an especially tight section, a green tree boa drops from its perch beside my head and knifes into the water.

Now the channel is dotted with floating mangoes. I grab one out of the water and hold it up like a trophy. Lukas cuts the engine and lets out a moan as we come upon the source of the fruit. A mango tree has fallen across the creek, completely blocking our way.

337

We jump out into knee-deep water and start to hack through the tree. It takes us two hours to cleave a space for our boat to pass. Just as we're pulling it through, Mawa whispers something and the Trios stop moving. Mawa's eyes are wide as he points at something over my shoulder.

Behind me, stretched out on one of the mango branches, a massive, green anaconda is sunning itself. Its body is more than five metres long and more than a foot thick. I stumble back toward the boat, shocked that this snake has been there the whole time. I expect the others to laugh at my fear but nobody laughs. Instead, everyone leaps into the boat and Lukas cranks the motor. At the sound of the engine, the snake slinks into the water, the yellow and black spots on its hide glistening as it disappears beneath the surface.

We spend the rest of the day crawling upriver and struggling through rapids. Our morale sinks as the motor worsens. I am beginning to think the trip is a mistake. We've made hardly any progress and we're wasting precious gasoline.

Finally, at six o'clock, we limp to shore, where the remains of a hunting camp are rotting among the trees. Ipiroke and I jump out with our gear and the others continue upriver to work on the motor and catch dinner. Twenty minutes later, a fire is roaring and we've hung our hammocks beneath makeshift palm-frond shelters. This camp has been used for centuries by Trio hunters, but Ipi says no one has slept here for at least five years. This is prime hunting territory, he says, but no one can afford the gasoline to get here.

The sun drops beneath the canopy and the jungle closes in. Ipi sings to the animals. He impersonates three species of monkey and four species of bird. As a cool evening mist descends on the river, a lone bird miles away responds to Ipi's call.

An hour later, the men return with a boatload of meat: Two twenty-kilogram tiger-fish, their striped flanks glowing orange in the dark, their mouths crammed with vicious teeth, and two twenty-pound *sipali*, electric stingrays, the lethal namesake of this river. Lukas hefts one of these from the boat and hands it to me, two fingers plunged through its eye sockets, its barbed tail hanging lifeless like a medieval mace.

Fireflies dance through the camp as Lukas cooks the fish in pepper water and Mawa smokes the remaining meat. We gorge ourselves on the rich, oily flesh. Then the Basia retrieves one of the detergent buckets. It is time to drink *casiri*.

Above us, the moon is reflected by the leaves, the green of the jungle giving way to the black and white of the night. My crew tell jokes I don't understand and laugh among themselves. The Basia lets loose a series of spectacular farts. Exhausted and drunk, we retreat to our beds.

I stumble into my hammock and the crossbeams of my shelter creak ominously. Then the whole structure collapses. I fall to the ground, the splintered timbers landing on top of me. My crew roars with laughter but within seconds they set to work. Lukas and the Basia disappear into the jungle while Mawa and Ipi cut new saplings and arrange them in a lean-to fashion. From somewhere in the darkness I hear ripping, and then Lukas and the Basia reappear carrying long strips of bark. They use the bark as rope, tying the new timbers together, and soon my shelter is rebuilt. I hang my hammock and test it out. The timbers are sturdy and will last another hundred years.

The men return to their hammocks and chat as a light rain trickles down from the canopy. I fall asleep to the whispers of their dying language.

The Trio have no words for measures of distance. Instead, distance is expressed in terms of the amount of time it takes to travel it. We have travelled one sunset so far. We have two more to go.

I roll out of my hammock at four thirty. It is dark and so time still sleeps. I am anxious to wake the crew and get on with the day. Our trouble with the motor coupled with unexpectedly shallow water has set us back at least six hours. If we want to make it to Sipaliwini tonight we're going to need at least eleven hours on the water.

Ipiroke is the first to rise. He emerges from beneath his palm-frond roof wearing only a blue Speedo-style swimsuit.

"*Andu!*" he yells, as he makes his way to the river by flashlight. "*Andu!*"

We are on the water by seven. The motor behaves itself and we let out a cheer. The river winds as if carved into the earth by a child. We pass a pile of boulders in the middle of the river and a colony of bats bursts out, enveloping the boat for a few seconds in a squeaking black cloud.

After we pull the boat through the first set of rapids, the sun finally breaks above the canopy and washes the river in light. To my left, a thick bamboo stand turns translucent in the sunshine, like the filaments of insect wings. We pass a *kankan* tree that leans over the river with the weight of more than fifty orapendula nests. Then Lukas cuts the engine and Ipiroke stands with his rifle. Mawa jumps out of the boat, wades to shore, and disappears into the bush.

Ipiroke aims at a branch fork near the top of the majestic tree and fires off a round. The gunshot echoes up and down the river and a huge mass of green plummets from the branch. It lands with a thud in the underbrush and Mawa yells. A chase ensues. From the boat, all we hear are Mawa's yelps and frantic footsteps. Suddenly, the injured animal crashes out of the bush – an iguana, nearly two metres long, desperate to save itself. Just as it stumbles to the water's edge, Mawa bursts out of the forest behind it and snatches it up by the tail.

He hands the writhing reptile to Ipiroke as Lukas starts the engine. Ipi digs the blade of his machete into the iguana's chest and I hear the whoosh of life escaping its body. He splits the animal open from its neck to its tail, its innards splashing into the water as he cuts them loose. In ten seconds the animal is empty, nothing but meat on bones, its lifeless fingers long and wrinkled, its gorgeous rack of blue-green spines slumped into its gaping body cavity. Ipi tosses the corpse into the hull and rinses his hands in the river. Lukas taps me on the back.

"*Ewana switi*," he says. Iguana tastes good.

We pass Agarapi Kreeki and the river narrows. Again we pull our boat up rapids, each of us up to our waists in rushing water, struggling to keep our footing. Lukas decides to motor up a particularly rough patch so we climb back in, and as the engine struggles against the current a metre-long *anyumara* – an oily river fish built like a tank – leaps into the boat. With a flash of steel, Mawa pins the fish to the hull with his machete. In seconds, its liver, intestines, and air bladder are floating past me.

This jungle teems with life. During the dry season, hardly a soul travels these waters, and even when the rains come it is rare for anyone to make the trip. Consequently, the region is stocked like a supermarket. We come upon a forest turtle swimming across the river. Lukas aims straight for it and Mawa scoops it out of the water and places it in the boat.

Walaba trees line the shoreline, their seed cases like thick boomerangs hanging from pieces of string. As we pass beneath a massive wasp nest, two metres long and thick as a tree trunk, Lukas yells something in Trio and turns the boat down a side creek.

Mawa and Ipiroke quickly reach for their rifles again. They load their guns as Lukas aims the boat at the shore and cuts the engine.

The boat scrapes up onto rocks and the men leap out. They scramble up a steep embankment into the bush and I scale the pile of equipment and follow them.

The jungle floor is still dark. The men race through the underbrush and I struggle to keep up. When I finally reach them, both men are whispering to each other and staring straight up into the canopy. A lump forms in my throat as I recognize the scene. The men are hunting monkey.

Forty metres above us, an adult red howler monkey sits with his back to a trunk. He does not move. He just stares down at us through the thick foliage. His bearded face is dark and ghoulish, his fur glimmering gold where it catches the rising sun. These are the monkeys who howl like banshees.

Mawa raises his gun but cannot get a clear shot. Lukas strips a young sapling and, hefting it like an axe, bashes it against the trunk of the monkey's tree. Now both men grunt loudly from deep in their throats, impersonating a predator or perhaps a jungle spirit. They are trying to scare their prey into moving, into exposing itself to Mawa's rifle.

In seconds, the monkey is leaping through the canopy. The men keep grunting as they track the animal. My training in primatology kicks in. I see flashes of red escaping to the west – two juvenile females, one juvenile male, an adult female with a baby on its back. I almost yell to the others but stop myself. Instead, I watch the monkeys disappear into the green, leaving their doomed patriarch behind.

I used to study monkeys. Now I'm hunting them.

A gunshot shatters the morning. I wait for a monkey to crash to the ground but nothing happens. Lukas and Mawa share a few words. Then Mawa leans his rifle against a rotting log, grips the blade of his machete between his teeth, and begins shimmying up a tree.

Far above, the monkey is slumped over, his tail wrapped around a thick branch to keep him from falling. Mawa glides up the trunk effortlessly. In less than a minute, he comes level with the dying animal, reaches out and unwraps its tail. The monkey slips from its perch, plummets through the branches, and lands at my feet with a thud. Its eyelids drift shut as it dies.

I drag the corpse back to the boat, gripping the base of its tail. Its crimson fur is softer than I'd expected.

In 1961, the American Baptist missionary Claude Leavitt arrived with his family in Trio territory. In the first Trio village he came to, the people told him they had seen only three *pananakiri*, outsiders, before him: Lodewijk Schmidt, a Surinamese explorer of African descent, and two Americans searching for a pilot named Redfern, whose airplane had crashed nearby.

Leavitt quickly realized his task of bringing the word of God to the Trio would be much easier if he could convince them to settle in a single village. Leavitt promised free health care and education in this new mega-village, and soon the Trio agreed to settle in an indigenous metropolis of sorts on the Wiumi River. The new village was called Alalaparu, which means "place of the brazil nut."

The Trio now had access to steel tools such as machetes, axes, and knives. A medical clinic was established and hygiene was improved, resulting in a greater life expectancy for the residents. A new airstrip reduced travel time to the towns on Suriname's coast from two weeks by river to two hours by air. The Missionary Aviation Fellowship began an air ambulance service, transporting seriously ill Trio to Paramaribo for treatment. Teachers were flown in to instruct the children and literacy flourished. A small store was set up, offering a range of material goods. Within a few years, the population at Alalaparu grew from one hundred to more than four hundred as Trio emerged from the surrounding jungles to take advantage of this newfound wealth.

But as is true of missionary work the world over, these real gains for the Trios were accompanied by an incessant evangelism that slowly overwhelmed them. There are pictures from the late 1960s of Granman Asongo in traditional Trio garb – his hair in a bowl-cut, his face painted for the hunt, wearing nothing but a loincloth. Soon after these pictures were taken, traditional Trio dress was outlawed and Asongo was sent to Texas for theological training – now he wears a suit and gold-rimmed glasses. The traditional system of government, in which a charismatic male elder oversaw a community held together by kinship ties and marriage, was replaced by hierarchical, centralized rule. Customary song and dance rituals were replaced by Christian music and ceremonies. Shamanism was forbidden, as was polygamy, body scarification, and the recounting of Trio mythology. All of these practices were made punishable by flogging. Perhaps the only social tradition to survive Leavitt's influence was the ritual drinking of *casiri*.

Alalaparu continued to grow and soon the population was hit by severe food shortages. With their new rifles, the Trio quickly exhausted the surrounding jungle of game. The Wiumi River ran dry for months at a time and fish populations dwindled. Gardens had been built too close together and a plague of ants decimated the crops. They were replanted too soon and the tired soil failed to produce.

Fifteen years after it was founded, Alalaparu was abandoned in 1975, the same year Suriname gained independence from the Netherlands. The community moved to virgin land on the Sipaliwini River. The new village was called Kwamalasamutu, or "place of sand and bamboo."

At five o'clock we come upon a small dugout canoe, large enough for a man and a hunting dog, pulled up the riverbank. Next to it, a fishing net is wrapped around a snag of wood. Ipiroke yells a Trio greeting into the bush but there is no answer. Then we round a corner and are blinded by the late afternoon sun. The jungle on the shore has vanished. We have entered the Sipaliwini savannah.

I stand up and peer over the mud banks and catch glimpses of a foreign landscape, miles of tall grasses and rolling hills dotted with solitary trees. It looks like the grasslands of Africa. The last five months in the jungles of Suriname have been a dark time for me. But now, as Lukas steers us down the middle of the narrowing river and the sky opens up, I feel that I can see beyond myself for the first time in a long time. The crush of sadness loosens its grip, slips away with the claustrophobia of all those trees.

An hour later we spot a triangular yellow flag fluttering over the river and Lukas cuts the engine. We've reached Sipaliwini. We dock beside an enormous boulder and unload our gear. Then Mawa leads us into the grasses. Two small children emerge from a copse of bushes and scurry ahead of us, eager to deliver the news of our arrival. We pass a family compound, rectangular thatched-roof buildings with walls of split bamboo. Soon there is more traffic on the trail, men wrapped in towels on their way to bathe, their muscular upper bodies decorated with green tattoos.

We stop at a small cooking shelter. Inside, two old Trio women tend a fire and spill the guts of fish into the embers. They wear only ancient, torn skirts. When they see me, they smile. Ipiroke drops the dead howler monkey at their feet and mumbles something to them. As we walk away, one of the women chops off the monkey's tail with a single swipe of her axe.

Mawa leads us to a modern-looking building and we climb the stairs to the second floor balcony. I open the door to one of the rooms and a horde of cockroaches scuttles to the corners. I set up my hammock in the semi-darkness. Beyond my window, the savannah stretches to the horizon, where a low range of hills glistens in the weak evening light.

Lukas cooks a late dinner of boiled iguana, rice, and pepper water. The meat is almost white and very tender. As I tear the flesh from its slippery hide, I chomp down on something metallic. I grimace in pain and spit the meat into my hand, assuming I've lost a filling. The men burst into laughter. In the middle of my palm is a small, grey gunshot pellet.

Sipaliwini wakes. The old women have started their fire and the roosters have begun to bawl. A small child, his legs crippled by some ferocious disease, crawls from a nearby hut and heaves himself into a hammock. Bees hum somewhere beneath our building and the village dogs emerge from the grasses. The sun is up and time has begun to flow again.

Before we leave, we drink *casiri* at the house of Mawa's friend. The brew is stronger than the batch we brought with us, and soon my vision is swimming. We load the boat, taking only the necessities and leaving the rest behind. Today, if we are lucky, we will reach the mountains of Mamia, the homeland of *okopipi*.

For some reason, a villager has given us a live boa constrictor in an old rice sack. The Basia is clutching a small birdcage containing a chestnut-bellied seedfinch, or *picollete*. As Lukas guns the motor, brown water sloshes past my feet; the poor turtle on a stick has shit during the night. Our boat is now a modest zoo.

Lukas decides to run the first set of rapids, the steepest ones we've seen besides Ewana Tepu. Like the rest of us, Lukas is very drunk. Halfway up, we hear a sickening sound – metal grinding on stone – and the boat lists and begins to drift backward. We jump out, the water up to our chests, and pull the boat to a small island of rocks to inspect the damage. Our prop is still there but one of the blades has been sheared off.

No one speaks. The Basia digs through our gear and pulls out a pot and the barrel of cassava bread. Inside the pot, chunks of boiled howler monkey sit in a congealed sauce. The men dig in but I wait for them to eat their fill. Then I reach reluctantly into the pot and pull out a severed hand, the fingers webbed with grease and onions. I nibble the primate palm, rip a piece of stringy flesh from it and feel revulsion rising from my stomach. I give the hand to the Basia, who tears the fingers off one by one and sucks the meat from the bone. I take a piece of breast instead, choking down my breakfast as the monkey fur tickles the back of my throat.

We spend the morning hauling our boat up rapids. The motor is almost useless now and we fight against the current. Kingfishers dive into the river beside us and emerge with wriggling fish. We pass beneath a dove's nest, the mother on top of her egg, waiting until the last moment to flee.

At noon we pull over, exhausted and baked by the sun. We are used to having the canopy to shade us but in the savannah the heat is relentless. The rainy season was due to start about three weeks ago.

I assume we've stopped to rest but Ipiroke grabs the axe and wades to shore. He scrambles up the bank to the base of a giant *awara* palm, the crown of which sags with three huge clusters of red fruit. Each cluster contains at least a hundred nuts, and this is what Ipi is after. But instead of shimmying up the trunk, Ipi begins chopping down the whole tree. Meanwhile, Mawa asks me for my lighter and I hand it to him without thinking. He wades to the other side and disappears into the grass.

Lukas and the Basia cast their fishing lines and I climb out of the boat. I watch Ipi work as I crouch in the shade beneath the river-bank. Then I hear crackling and smell smoke. I climb up to the grasses and am almost engulfed by flames. Mawa has set the savannah on fire.

I can see him a half mile away, holding my lighter to a swath of grass as walls of flame sweep between us. I scramble back down to the shore but Lukas has moved the boat upriver and there's no way to escape. The crackling gets louder and the air becomes an oven. The grasses above me shrivel and burn and I prepare to leap into the river.

Somehow, the fire fails to catch the trees at the water's edge and burns itself out. Lukas and the Basia are laughing at me and Ipi continues to hack at the palm. I wait a few minutes and then climb back up.

I wander through a scorched landscape. Clouds of thick smoke drift on the breeze and the charred earth melts the soles of my flip-flops and bakes the soles of my feet. In the distance, the fires continue to race as Mawa walks among them, his head bent low in search of something. Lukas has joined him now and is setting his own swaths of savannah ablaze.

Then a cry goes up from the river and the palm begins to fall. It drops slowly at first, a shower of nuts preceding it to the ground. Then it disappears below the riverbank and lands with a violent crash. Ipiroke miscalculated. The tree now lies across the river, barring our return journey.

I climb back down and splash into the water. The Basia drifts the boat back and we fill the hull with nuts. Mawa and Lukas appear at the top of the riverbank, covered in black ash. They are carrying six savannah turtles, flushed out of hiding by the flames, sticks thrust through their arm holes, twenty-four red-soled feet flailing. The Trio treasure these turtles as pets. These are gifts for Granman Asongo.

We continue south, and by four o'clock, the creek is no more than a foot deep. Lukas has cut the motor and Mawa and Ipiroke use sticks to punt the boat. Then, from a dark cluster of trees on the riverbank, a lone man emerges and waves us down. This must be Winni. We have arrived in Mamia.

Lukas beaches the boat and we carry our gear up a small rise, where a firepit blazes and a makeshift hammock camp leans precariously. Winni eyes me and then grabs my hand and shakes it. He is perhaps forty-five years old with a slight build, weary eyes, and the ubiquitous bead jewellery of the Trio around his neck.

Winni's daughter, Yana, is here as well, a beautiful teenager with a glossy black ponytail. As we reinforce the shelter and cover it with our tarps, Yana butchers an iguana. She holds it by the tail and sweeps her machete across its belly as if playing the cello.

349

Winni asks to speak with me. He has changed into a white button-up shirt and black pants. Even out here, the greeting of a stranger is a ceremony. Ipiroke translates from Trio to Sranantongo for me. When I first arrived in this country, I needed translations from Sranantongo to English.

Winni welcomes me to the Sipaliwini Nature Reserve. Granman Asongo had radioed him to tell of the white man and his search for *okopipi*, so he and his daughter hiked four hours this morning to meet us. Winni will be our guide into the mountains. He is happy I am here and hopes that I will pay him. He also hopes I find what I'm looking for, because without the rains there are no guarantees. He finishes by warning me not to leave Mamia with *okopipi* in my backpack.

I thank Winni, give him the money I'd brought him, and promise not to take any frogs. He seems to relax a little and we settle in for dinner. The others have caught a few *dwala* and have boiled the small fish with the iguana. As we eat, the sun sets and the *mopira* flies ease. Afterwards, we hang our hammocks and my companions collapse into them, exhausted and happy to have arrived.

Before bed, I light a smoke and walk out to the savannah. The moon is full and casts a cool, grey light. To the south, the Four Brothers of Mamia rise up, Suriname's last sigh before Brazil. Mosquitoes have replaced the *mopira* and fireflies dance among the grasses. Forty years ago, another white man came here to climb the Brothers. He fell off the mountainside and died before help could reach him. Amasina was on that trip, just a young boy back then, a child shaman eager to visit the spirit realms. He was powerless to save that man. The missionaries had just arrived.

Today, the Trio face a new threat, a challenge more modern yet potentially as disruptive as a Baptist mission. The government, its eyes firmly fixed on the rich deposits of bauxite and gold that are believed to exist in the southern savannah, refuses to give the Trio legal title to their lands. The reason, of course, is that the mines in the north are nearing exhaustion. When the extraction industry casts its inevitable gaze southward, the government wants to ensure that the Trio will be powerless to oppose it.

The Trio are taking action. In partnership with the Amazon Conservation Team, they are creating the first GPS resource maps of their territory, identifying everything from hunting spots, fishing spots, and garden plots to old village sites and cemeteries. These maps are crucial to demonstrate their long-standing use of the land – that these lands are not, as corporate and government officials would like us to believe, empty. The Trio are also actively courting projects like the one at Kwamala, in the hopes that tourism can outcompete the potential of mining.

Our search begins with a prayer. We gather in the dark at the edge of camp, where the jungle gives way to savannah, where the trail to Brazil begins. Winni asks us to bow our heads. He speaks quickly, efficiently, reverently – the voice of a man who knows faith and timing are of equal importance when you live in the bush. He uses a mixture of Trio and Sranantongo. I pick out mentions of God, happiness, white men. As he speaks, grasses squeak in the morning breeze and the sun still sleeps behind the Mamia Mountains.

We begin to walk. Winni and his daughter take the lead and I follow them closely. After three long days pulling our boat upriver, hiking feels like flying. But Winni goes slowly, bent at the waist, sweeping his machete across the grass in front of him to scare off snakes. He wears ancient rubber boots and Yana wears flip-flops. She is thirteen and has lived her whole life in the savannah.

The footing is bad, made worse by the dark. The trail is nothing more than a narrow break in the grass, worn into the land by the family of two in front of me, the only people who walk here. Their home is on the far side of Mamia, a shack on an airstrip that no one uses. Their lives are defined by the scars they leave on the land – this near-invisible trail, the blood-red machete wounds on trunks of trees, a pile of charred chicken bones. It is a reciprocal thing, this wounding and scarring. A country leaves marks on its people just the same.

After half an hour we stop on a small rise. The young sky turns red then orange. When the first rays of sunlight sneak out from behind the Four Brothers, Yana giggles.

"*Okopipi drape*," says Winni, pointing toward a valley between the mountains. This is where the blue frog lives, the only place on earth.

Now the footing is treacherous. Without a trail we stumble over thick tussocks of grass and sharp rocks the colour of rust. At any moment I could step on a snake and it would all be over. Winni and his daughter pull ahead as my ankles twist and swell. The others take their time, trade jokes and scan the horizon for game.

We outrun the sun to the foot of the mountains, where Winni stops in the shade and considers our next move. He confers with his daughter, who points toward a patch of giant heliconia palms, a sinister mass of green at the bottom of the valley. Winni thinks for a moment, then nods and heads straight for it.

Among the palms the air is cooler, the ground black mud. But everything is closer here, darker, and I soon find it difficult to breathe. My boots sink to the ankles, but Winni's daughter seems to float. We find a small, stagnant creek and follow it up the valley, bracing ourselves against the heliconia that grow along its banks. Soon the water is gone and we hike up the middle of the dry creekbed.

And suddenly we are searching. Without a word, Winni has stopped and begun examining the crumbling banks of the old creek, and everyone behind me has fanned out in twos and threes. We are in thick jungle now, the insects singing, the lizards scrambling for cover. The little water that remains from the last rainy season glistens in the mud where the sun pokes through the canopy.

I stay close to Winni. He goes up one side of the creekbed and I go up the other. I look inside every hole and behind every rock. I lift dead cecropia leaves from the forest floor and flip old logs to peer beneath them. We go softly, quietly, as if meditating. The footfalls of my crew are above us now, the men exploring the steep hillsides.

After a half-hour, Winni turns to me, his face blank but for a slight frown. "*Watra tumsi saka,*" he says. The water is too low. The same problem that plagued our entire journey – the lack of rain that made the Sipaliwini rapids impassable – now threatens our final goal. Winni has never had to search this long before.

The crew keep looking but I take a break. I lean up against one of the boulders that, when the rains finally come, will give this stream its shape. I realize I may never find what I'm looking for; the exhaustion that's been building for so long finally descends. I shut my eyes and listen as the Indians scour the bush, the soft roar of the rainforest rising and falling like waves on a distant sea.

When I last spoke with Emma, all those months ago on the Stadszending phone, she accused me of being a hopeless romantic. Looking back, this is what upset me most. She made me feel like a naive young man who secretly believes in the myth of Eden, the myth of the Noble Savage. But I am not a hopeless romantic. I am, instead, a very hopeful one. I love those old stories not because I believe they are true but because of their power, the bearings they provide, the wisdom at their core. Those stories give us the wisdom of hope – for us, for the animals, for the forests, for the earth, traction up the mountain at a time when hope seems slim. This is why I gave up studying monkeys all those years ago – given a choice, I prefer the illumination of stories to the rationalism of science.

But now, utterly worn out and slumped against this stone, I finally fall prey to that nefarious force I've been struggling with ever since I left Suriname the first time around, that mistrust of wonder my own culture has bred into me. I begin to feel completely ridiculous. Has five months of searching really come down to this, a heartbroken white man in an ancient forest with nothing at all to show for it? Has my *koiri* simply been futile? Because I've been stumbling after Eden for a long time now, and all I've done is trade one paradise for another. I've lost Emma for good. I know this in my bones. And with the narcissism of a young writer, I've drawn parallels between my own little heartbreak and the monumental struggles of a beautiful, bountiful nation. I might be Werner Herzog, my steamship slipping from my back. I might be the Conquistador of the Useless.

This hidden garden is under siege, and I blame everyone. The Baptists, the Surinamese government, the mining and timber industries, my own monolithic society, the short-sightedness of the Trio themselves. I blame Emma for losing hope; I blame myself for the exact same thing.

And then someone screams. A young woman's voice, halfway up the eastern hillside, crackling with fear or excitement or both. Then the sound of thumping footsteps, more nervous voices. I push off from the rock and scramble up the steep riverbank, *maka* spines stabbing my hands, my backpack catching on low-hanging vines that shower me with ants. Winni races past, unsure of what his daughter has found. I do everything I can to keep up.

The others have gathered around a small palm tree. At its base, the leaves and old nuts have been pushed aside, and in the middle of this clearing sits a tiny blue frog. Its iridescence is stunning against the mud and decaying leaves. Its legs are a dark cobalt and its bright blue back is mottled with deep blue spots. I inch closer.

I can't believe what I'm seeing. Then Winni puts his arm out to stop me, leans down, and gently plucks *okopipi* from the ground.

He holds the frog out to me. I hadn't planned on touching it – its poison can kill a man, and the oils in my palm might harm it – but I can't help myself. It weighs almost nothing. It is less than three centimetres long. And as I feel its frantic heart beat through its cold, wet skin, I experience a moment of sheer and shaking wonder unlike anything I've ever felt before. My guilty hand from Raleighvallen is no longer guilty, but blessed. Nature is still here, asking for protection – we haven't lost her yet. I am holding the quintessential spirit of Suriname, the soul of the Last Eden, in the palm of my hand.

THE VOYAGE OF AITKANTI

The trip home to Kwamala is arduous even though the current is with us. We set more of the savannah on fire and chop our way through the *awara* tree. We shoot rapids we shouldn't and almost capsize three times. We happen upon a school of *karari* floundering in a rocky pool and slaughter more than sixty of them, poking sticks through their gills and arranging them in neat sets of ten in the bow. Meanwhile, a nasty blister opens up on my left buttock.

We arrive in Kwamala to little fanfare. As we unload our gear a young Trio girl watches from a submerged rock in the middle of the river. She waves to us and steps into the water, slipping beneath the surface without a splash.

Granman Asongo gives me a bear hug when he sees me. He is happy we are safe and that no one died. We present him with the turtles, and he asks if we found *okopipi*. I tell him yes. He asks if I brought the frogs back with me and I laugh and tell him no.

I find Bruce in his hut by the airstrip. He's conducting an interview with one of the oldest women in the village. The interview has lasted for three days. I listen in as she describes what one of the plants on Bruce's list is used for. Bruce looks like he hasn't slept in weeks. In the distance, a plane approaches from the city.

Amasina appears in the doorway. He hands me a chunk of smoked tapir meat wrapped in newspaper and asks me to write the name "Joseph" on the outside. As I do this, he argues with the old woman about the use of one of Bruce's plants. He says it's good for sore throats; she says it's only good for building houses. The plane arrives. Amasina leaves the hut with his package of meat and hands it to the pilot. Moments later, the plane taxis to the end of the runway and speeds toward the shaman. Just as it seems it might never lift off, the plane rises with a howl and heads for the coast.

The next day, as I wait for another plane, I join Lukas, Ipiroke, and Mawa, who are inside an abandoned cookhouse, sitting around a half-empty bucket of *casiri*. They are celebrating our successful return, but they are also mourning the end of the adventure and their return to unemployment.

I sit and drink with them. Then I bring out my camera and show them the photos from our journey. They watch as we drag our boat up the Sipaliwini River, bunk in hammock camps centuries old, and walk behind Winni and his daughter. But then the photos go dark and lose focus. Indian faces, flashes of jungle, men with a blur of blue in their hands. I click forward for the photo of me with *okopipi*, the blue jewel of the jungle.

There isn't one.

Four months before I arrived in Suriname, a female leatherback turtle named Aitkanti left here on an ocean voyage. Before she left, researchers from the World Wildlife Fund tagged her with a radio transmitter, and during my travels here I have kept tabs on Aitkanti's *koiri* through the satellite map on the WWF's website.

The beaches of Galibi, at the mouth of the Marowijne River, are the biggest Caribbean nesting grounds for the leatherback turtle. Leatherbacks are true travellers – in her lifetime, a single female will swim tens of thousands of kilometres, from cold water feeding grounds in the North Atlantic, where she gorges herself on jellyfish, to tropical beaches where she lays her eggs. Though their migrations are still poorly understood, we do know leatherbacks usually return to their home beaches every year to lay their eggs, in the same sands where they themselves hatched.

Over the last five months, countless people have told me to visit the beaches of Galibi. It must be a remarkable sight – hundreds of lumbering, exhausted, prehistoric giants dragging themselves ashore at night, digging holes in the sand with their forelimbs and dropping their eggs like stones. Unfortunately, I have neither the money nor the strength to remain in Suriname until April, when breeding season begins. There is only one more place I want to visit in this country, and then I'm going home.

My plane lands at Raleighvallen like a kid on a trampoline. Biggaman, Captain, and Doctor Five help me unload next to the old wooden sign that trumpets STINASU's motto: *Natuurbehoud is Zelfbehoud*. Nature preservation is self-preservation. The men ask where I've been; I tell them but they don't believe me. Then I search Foengoe Island for Petrus but can't find him, so I shoulder my backpack and walk down to the water.

Now the rainy season has arrived, the Coppename River has flooded. Only the tallest boulders are visible in the rapids. Young cecropia leaves quiver on the surface, while below, *anyumara* struggle up the whitewater. Piranha, *tukunari*, caiman, electric eels – all have returned in a reunion of water and spirit. The river now reaches halfway up the hillside, halfway to the foot of the new visitor's centre. Perhaps four months of travel have tempered my idealism (or maybe it's just the swollen river), but the building seems smaller now that it's almost complete, less imposing and in perfect proportion to the rainforest surrounding it. I'm struck by the surprisingly rational thought that ecotourism, despite all of its contradictions, might be our best shot at preserving the jungles of Suriname after all.

I find Set-man among the boats. In minutes we are motoring upriver. We zigzag back and forth, from bank to bank, the stones submerged but just as dangerous, their coordinates burned into Set-man's memory. He lets me out at the beach where I lived my last year as a scientist, where my journey began all those months ago. Instead of visiting the house, I join the trailhead and hike into the jungle.

For the next two hours I wander the old trails, the subtle pathways through the rainforest that followed the overhead routes of my monkeys. The memories come flooding back. The herd of *pingos* I surprised up on Bonkistraat, hundreds of foul-smelling boar rummaging through the mud and clacking their tusks at me in warning. The harpy eagle attack down on Avenida Maravillosa, that awful morning when the lights went out, the massive bird blotting out the sun as she descended on my troop. The ferocious storm I sat through on East Loop, perched on a camp chair, the monkeys taking cover, a cathedral of ancient trees collapsing in the gale. The rainbow boa we found one night on Rock Trail, wrapped around the base of a stone like a haute couture accessory. The tapirs I spooked, the killer bees on wing, the giant river otters in Manari Creek who howled like banshees. The rumbling armadillos, the coiled bushmasters, the sage-like sakis, the anxious agoutis. The rare family of bush dogs I stumbled upon. And of course, the jaguar who coughed, the cat we heard but never saw. The one who chased my boss and me back to camp one day. And the snake that lay in wait just steps from our house – the notorious fer-de-lance, coiled to strike in the middle of the trail.

I reach the heart of the study site. Bathed in sweat, I cock my head back and howl.

"Whoo!"

The forest swallows my voice. Then a faint echo, muted by the humid air.

"Whoo! Whoo!"

Two howls means you've got monkeys.

I find them at the top of Parbolaan – John and Andy, the young Americans living my old life. The canopy drips with restless capuchins.

"Starbucks?" I ask.

"Yup," says John. "They're heading for the bamboo."

I run beneath them, frantic to see one I recognize before they disappear. I try to remember what Banana, Russell, Mignon, and Suri Rama look like; I remember the jungle but not my monkeys' faces. One by one they drop from the trees and sweep into the bamboo, and one by one I fail to identify them. The last to vanish is Bruce, John tells me, the massive alpha, who ousted Jules in a now-legendary battle. I knew Bruce back when he was leading another troop. Now he's a stranger to me.

We spend the rest of the afternoon circling the bamboo, waiting for the troop to emerge. At six o'clock, exhausted and monkey-less, we head home. Over dinner, I tell the crew my plans for tomorrow night, the real reason I've returned to Raleighvallen. They agree to take the day off and join me.

We leave the trailhead at noon, weighed down with food, water, and mattresses tied to our backs. We pass familiar landmarks: the massive boulder called Harrison Ford Stone, which resembles the enormous boulder from Indiana Jones; *Fisi Ston*, the wall of water-worn rock that looks like the gills of a fish; the ominous stand of heliconia plants, their leaves giant parasols shading the trail. Three hours later we reach the sun-baked West Platte, a vast slab of deserted granite. From here we catch our first glimpse of the mountain, the Voltzberg Dome, rising 250 metres above the canopy like a crouching elephant. On the other side of the Platte we pass through Russell Mittermeier's old research camp, where he first fell in love with

Raleighvallen in the 1970s, decades before he led the effort to save this country's rainforest.

The trail takes us to the foot of the mountain, where we stop for water, adjust our packs, and steel ourselves for the ascent. Then one by one, we pull ourselves up into a narrow fissure and scramble up the wall. After ten minutes of heavy going, we leave the fissure and emerge from the slim band of trees encircling the mountain. Now there is nothing but sheer rock, pounding sunshine, and occasional cacti clutched to the slope. For the next thirty minutes, we trudge up the mountain in a long line. The higher we go, the steeper it gets. We lean into the incline as far as we can, our legs burning, our hands on hot stone.

With one last push, we stagger onto the summit. We drop our bags, peel off our sweat-soaked clothes, and lay them out to bake in the sun. We celebrate in our underwear as all around us the unbroken rainforest of Suriname rolls to the horizon. We dance among the cairns of rocks erected by previous climbers, the soles of our feet ablaze, leaning into the breeze blowing north to the sea. We see France, we see Guyana, we imagine we can see Brazil. I've climbed this mountain countless times but the view always leaves me speechless.

I think of Voltaire's *Candide*. Suriname is where Candide finally renounced his optimism, after experiencing all the miseries known to man, but his last words, at the end of his travels, suggest even he saw an encouraging way forward. "Let us cultivate our garden," he said.

I think of Robin "Dobru" Raveles, Suriname's great nationalist poet, and his most famous poem, "*Wan Bon* (One Tree)":

Wan bon	One tree
someni wiwiri	so many leaves
wan bon	one tree

wan liba	one river
someni kriki	so many creeks
ale e go na wan se	all are going to one sea
wan ede	one head
someni prakseri	so many thoughts
prakseri pe wan boen	thoughts among which
moes de	one must be good
wan Gado	one God
someni fasi foe anbegi	so many ways of worshipping
ma wan Papa	but one Father
wan Sranan	one Suriname
someni wiwiri	so many hair types
someni skin	so many skin colours
someni Tongo	so many tongues
wan pipel	one people

We watch the sunset in silence, the endless jungle turning pink then purple, the daylight slowly dying. Then we lie down on our backs, drink *palum*, and share stories about our monkeys. The stars roll behind storm clouds. The rains begin and the wind howls. We build a fortress of stones and fall asleep beneath it.

The Voyage of Aitkanti is the ocean-going version of that story taught in geography class, a story so astonishing it reads like a myth – two hundred million years ago, all the land on earth was con- nected. A super-continent called Pangaea was surrounded by a massive ocean called Panthalassa. During the Jurassic Period,

Pangaea broke up into two smaller continents, Laurasia to the north and Gondwana to the south.

Even to adult ears, the fantastical names of Pangaea, Panthalassa, Laurasia, and Gondwana give this story the air of a legend, the qualities of a creation myth. But it is not a myth. It is a story with a wisdom science has proven to be true, a wisdom any child can demonstrate with a piece of cardboard and a pair of scissors. All the earth used to be connected. Like a giant jigsaw puzzle, the continent of South America fits perfectly into the Atlantic curve of Africa. Two hundred million years ago, the jungles of Suriname and the jungles of the Slave Coast were exactly the same jungles.

Two days ago, just back from Kwamalasamutu, I logged into Aitkanti's website to check on her progress. She was more than four thousand kilometres away and diving deep, a breath of hope bound for the land of the ancestors, somewhere off the coast of West Africa.

I wake at five o'clock, lift the tarp and roll out into the night. The air is damp, the stone is cold, and the jungle below us is quiet. I flip on my headlamp. The darkness explodes into white. The breath of the rainforest has risen this morning, shrouding the Voltzberg in fog.

I have a flight to catch and John has monkeys to study, so we let the others sleep. Shivering, dressed in layers, we leave the summit, the way forward dimly revealed by our headlamps. The slope is nearly vertical and perilously slick. We go slowly, blindly, often on our bums. There is no trail, just sheer rock in thick darkness, and we get lost a number of times. We might be the only waking humans for hundreds of kilometres. We might lose our footing at any moment and plummet to the forest below.

Halfway down, we enter the belt of trees that encircles the dome and drop below the fog. The air warms as we follow a smooth cleft in the stone, worn into the mountain by centuries of run-off rain.

The last section is the trickiest, a steep descent down a fissure lined with pineapple plants and thorny vines. The skin on my hand tears open as I slip and reach out to catch myself. The nocturnal sounds of the jungle, the nighttime riot of the rainforest, slowly rise to our ears.

Finally, we sink into the loamy floor of the jungle. John radios the summit to let the others know we made it down safely, and then we search for the slim trail that led us here. My skin crawls as we enter the looming trees. All of the hunters are awake right now and my headlamp is slowly dying.

John leads through the darkness. We should be talking, or whistling, or shouting, but for some reason we're silent. Occasional eyes glow red in my fading light. An eerie thumping rises on the air. I struggle to stay calm, to stifle my fear, to remind myself Werner Herzog was wrong. The jungle was not created in anger. The riverbones are an exception, not the rule. The harmony of the rainforest is not one of obscenity and murder but of exuberance, benevolence, and peace.

Then John screams and stops dead. I slam into him. *Shit*, he says. *Shit. Shit!* The hairs prickle on the back of my neck. I peer over his shoulder to see what he's found.

She is just in front of us, beautifully camouflaged in a weak halo of light. A coiled fer-de-lance, South America's deadliest snake. We shake branches to scare her off. We stomp our feet but the snake doesn't move. She just sits there in strike position, stubborn and unyielding, the ideal conservationist. Convinced of her power to defend life itself.

ACKNOWLEDGEMENTS

A special thank you to my friend Ric Arseneau, who so generously plucked me from his garden one summer and sent me on my way; to Jason Rothe (www.jasonrothe.com), my trusty photographer and road-warrior – here's to many more trips together into the unknown; to my sister, Kathryn, and my sister-in-law, Andrea, who fell in love while I was gone; and to Samantha, my *lobiwan* and closest friend, who makes me feel like the luckiest monkey around.

A traveller is nothing without his foreign hosts, collaborators, and friends, so a big thank you to Steven Schet, Natascha Veninga, Monique Pool, Chuck Hutchinson, Bigga, Doctor Five and the Raleigh Boyz, the Aboikoni family of Saramaka, all my mates in SUR10/11 of the United States Peace Corps, Paul Woei, Bruce Hoffman, Paramount Chief Granman Asongo, Lukas, Mawa, Ipiroke, the Basia, Winni and all my Trio friends, and, of course, Dr. Sue Boinski, who first sent me to Suriname all those years ago.

On the writing and research side of things, sincere thanks to the following for their insight, support, laughter, and red ink: my superb agents, Anne McDermid and Martha Magor, my tag-team of exceptional editors, Chris Bucci and Dinah Forbes, my publicist Josh Glover, Ian Pearson, Moira Farr, Rosemary Sullivan, the Banff Literary Journalism crew, James Little, Bill Reynolds, Chris Tenove, Fergus MacKay at the Forest Peoples Programme, Dr. David Hammond, Richard Poplak, Giles Blunt, Jonathan Hayes, Mitch Kowalski and everyone at the Toronto Writers' Centre, Randy Krumme, Stephen Morris, Becky Toyne, Mark Berube, Alistair Stewart, Sheila Koffman and the gang at Another Story Bookshop, and everyone at Tinto Coffee House. A very special thanks to two unforgettable mentors, Carolyn Smart and Andreas Schroeder, for their early inspiration and encouragement. Thanks also to *explore* and *The Walrus*, where excerpts of this book originally appeared, and to the Toronto Arts Council for financial assistance.

Finally, to Bodi, wherever you are: *Waka bun.*